106

2007
70

NONLINEAR VIBRATIONS

PURE AND APPLIED MATHEMATICS
A Series of Texts and Monographs

Edited by: H. BOHR • R. COURANT • J. J. STOKER

VOLUME II

NONLINEAR VIBRATIONS
in Mechanical and Electrical Systems

J. J. STOKER

INSTITUTE FOR MATHEMATICS AND MECHANICS,
NEW YORK UNIVERSITY, NEW YORK

INTERSCIENCE PUBLISHERS, INC., NEW YORK

INTERSCIENCE PUBLISHERS LTD., LONDON

INTERSCIENCE PUBLISHERS, Inc.
215 Fourth Avenue   New York 3, N. Y.

For Great Britain and Northern Ireland:
INTERSCIENCE, PUBLISHERS Ltd.
2a Southampton Row   London W.C. 1

PRINTED IN THE UNITED STATES OF AMERICA
BY WAVERLY PRESS, BALTIMORE, MARYLAND

To

R. Courant

Introduction

During the past twenty years the nonlinear problems of mechanics and mathematical physics have been vigorously attacked by scientific workers having a wide variety of aims and purposes: one need only recall, for example, the advances made in fluid dynamics—particularly in gas dynamics—and in plasticity and nonlinear elasticity. The purpose of the present book is to give an account of another small segment of the general field of nonlinear mechanics, i.e., the field of nonlinear vibrations. Actually, the title of the book is rather too inclusive since only systems with one degree of freedom are treated, but this is rather natural because of the fact that more general systems have been studied very little.

It is perhaps worth while to consider for a moment the reasons why one should be interested particularly in the nonlinear problems of mechanics. Basically the reason is, of course, that practically all of the problems in mechanics simply *are* nonlinear from the outset, and the linearizations commonly practiced are an approximating device which is often simply a confession of defeat in the face of the challenge presented by the nonlinear problems as such. It should be added at once that such linearizations as a means of approximation have been and always will be valuable—in fact completely sufficient—for many purposes. However, there are also many cases in which linear treatments are not sufficient. For example, if the oscillations of an elastic system result in amplitudes which are not very small, then the linear treatment may be simply too inaccurate for the purposes in view. In such cases the accuracy can often be improved sufficiently by carrying out further approximations of the same sort as are involved in linearizations. However, it happens frequently that *essentially new phenomena* occur in nonlinear systems which can not in principle occur in linear systems. A familiar example of such an essentially nonlinear phenomenon in the field of gas dynamics is the building up of a discontinuous shock wave from a smooth wave. In nonlinear vibrations examples of the same sort are the occurrence of *subharmonic* forced oscillations in a wide variety of systems, the occurrence of systems (the so-called self-excited systems) in which a *unique* periodic free oscillation occurs, and the

occurrence of what are called *combination tones*.   In this book, the
principal aim is not so much to introduce methods of improving the
accuracy obtainable by linearization, but rather to focus the attention
sharply on those features of the problems in which the nonlinearity
results in distinctive new phenomena.   One of the most attractive
features of the subject of nonlinear vibrations is the existence of a
surprisingly wide variety of such distinctive new phenomena; and,
what is perhaps still more surprising, these phenomena can be treated
by methods which are interesting and instructive in themselves
without being difficult, and which do not require the use of sophisti-
cated mathematics.

This book has been written with the needs and interests of several
classes of readers in mind.   To begin with, the author has wished to
present the underlying principles and theory in such a way that they
can be easily understood by engineers and physicists whose primary
interest is in applying the ideas and methods to concrete physical
problems.   The author has the impression that the existing fund of
knowledge in nonlinear vibration theory has not been used in practice
as much as it could and would be used if engineers and physicists
were more familiar with it.   In view of the author's own general
interests—not to say biases—it would be strange if the needs and
interests of applied mathematicians were to be neglected.   For this
class of readers the author has emphasized the various known types
of physical problems in the field which lead up to the questions of
mathematical interest, and on the other hand carried out detailed
treatments, particularly in the Appendices, of a variety of important
problems of a mathematical character, some of which constitute
results achieved only in the last few years.   Thus the book might be
hoped to serve those readers who wish to be brought up to the border-
line where new discoveries are being made.   At the same time these
readers are supplied both implicitly and explicitly with hints regarding new problems to be tackled and with a number of ideas and
methods that could perhaps be used to solve them.   The author's
acquaintance with the material of this book arose through seminars
and lecture courses conducted from time to time at New York
University over a period of nearly ten years.   The subject has
invariably proved to be interesting and stimulating to the students.
Consequently the author hopes that the book may prove useful to
other colleagues in the teaching profession who may wish to conduct

similar seminar or lecture courses in nonlinear vibrations, or who may wish to supplement a course—even a quite elementary course—on ordinary differential equations with some of the striking illustrative material which occurs in such profusion in the field of nonlinear vibrations.

In attempting to please too many classes of readers the author incurs the well-known risk of pleasing none of them. In the present case, however, this risk is to a large extent obviated by the character of the material itself. While the main types of problems can be treated rather satisfactorily with the knowledge and use of little more mathematics than elementary differential equations, the problems, if investigated thoroughly and deeply, lead at once to questions of great subtlety and interest from the mathematical point of view—a case in point is the so-called "difficulty with the small divisors" which occurs as soon as one treats the problem of combination tones.

Before outlining more precisely the actual contents of the book, it should be said that the book is not intended to furnish a complete survey of the more recent literature in the field. This is particularly true with regard to the Russian literature, which is well known to be important and extensive. Fortunately, the recent book by N. Minorsky (*Introduction to Non-linear Mechanics*), and the translations by S. Lefschetz of books by Kryloff and Bogoliuboff and by Andronow and Chaikin, go far toward making this literature available. One omission which the author regrets is the theory of Liapounoff for the discussion of stability questions; but to deal adequately with this theory requires more space than would be reasonable in a book like the present one.

The source of practically all the basic mathematical ideas and also the techniques in nonlinear vibrations is the work of Poincaré, while the specific basic physical problems treated at present in nonlinear vibrations were introduced by Rayleigh, van der Pol, and Duffing. The object of the present book is to give a connected and systematic account of this work, which includes most of what had been done up to about 1930. The author would not presume to give a survey of outstanding work done after 1930 in the field. Nevertheless, a considerable amount of quite recent work is discussed in the book: the work of Levinson and Smith on the existence and uniqueness of the periodic solution in a very general case of the self-excited type, and the quite recent work of Haag and Dorodnitsyn on asymptotic developments for the period and other quantities associated with relaxation oscillations, are examples in point.

There are six chapters in the main body of the book, and six Appendices. The first five chapters of the book are elementary in character, the sixth chapter is rather less elementary, while the Appendices are on the whole not elementary, containing, as they do, rigorous existence and uniqueness proofs.

Chapter I consists of a short, though fairly complete, summary of the theory of linear vibrations for a system of one degree of freedom having constant characteristics. This chapter serves both for reference and for contrast with the results of the nonlinear theory. Chapter II treats easily integrable nonlinear systems in which no external forces depending on the time occur. A considerable number of physical problems illustrating the material is given, and a first glimpse of the advantages to be gained by working geometrically in the phase plane is achieved. In Chapter III free oscillation problems of a type like those of Chapter II, but not so easily integrable, are studied in detail by working in the phase plane, again with reference to a variety of physical problems. In the course of this chapter the graphical method of Liénard is introduced and applied, the theory due to Poincaré of singularities of first order differential equations is developed in order to obtain the criteria for their classification into types, and the notion of the index of a singularity is introduced. Finally, the usefulness of these ideas is illustrated by solving concretely a number of physical problems. One of these is a problem in elastic stability, which is treated dynamically. Another is the interesting problem of the pull-out torques of a synchronous motor, which is given a detailed treatment.

In Chapter IV problems of nonlinear *forced* oscillations are taken up and specialized to the cases in which the nonlinearity is provided solely by the "elastic" restoring force. This is the type of problem first treated with significant results by Duffing. Such problems arise any time that a system with elastic restoring forces is subjected to periodic external forces which cause sufficiently large amplitudes. The "response curves" are first studied—that is, the curves showing the amplitude of the forced oscillation as a function of the frequency for a given amplitude of the external force. In this connection the problem of hunting of a synchronous motor is considered. A curious "jump phenomenon" is studied which has often been observed experimentally. The effect of viscous damping is considered. The occurrence of subharmonic oscillations is discussed. The problem of the occurrence of combination tones is considered, and the general problem of the stability of the periodic oscillations is formulated. However, this chapter has another purpose aside from presenting the solutions to certain specific problems, and which is perhaps even more important. That purpose is the presentation and discussion of a considerable *variety of analytic methods* useful in treating the periodic solutions. In particular, iteration and perturbation methods are explained and applied to obtain directly the solutions of the differential equations; both methods are also applied more indirectly as a means of determining the coefficients of the Fourier series developments of the solutions. Thus the same problem is sometimes treated a number of times by different methods in this chapter. Special mention might be made of the iteration method of Rauscher, which is also explained in this chapter. At the end of the chapter a table (similar to one in the little book of Duffing) is given in which the contrasts between linear and nonlinear systems are pointed out. In the more recent literature on nonlinear vibrations the problems of the type discussed in this chapter are usually treated rather summarily, if at all; consequently it was felt that a detailed treatment might be found useful.

Chapter V is devoted entirely to problems in which the non-linearity occurs in the "damping" terms (i.e., the terms depending on the velocity, rather than on the displacement) in such a way as to cause what are called self-excited or self-sustained oscillations. Systems of this kind are very common in nature: they occur always in fact when a periodic motion is maintained through absorption of energy from a constant flow of energy. The best known and most important technical applications occur in electrical systems containing vacuum tubes, in which the energy for the oscillations is supplied by a direct current source. Oscillations of the same type occur frequently also in mechanical and acoustical systems. In fact, Rayleigh probably pointed out the first example of the sort in the case of the production of a sustained note from a violin string caused by bowing it. The failure of the Tacoma bridge a few years ago is generally ascribed to a particularly heavy self-excited oscillation in which the constant energy source was the wind. The flutter of airplane wings is another example of the same kind.

Chapter V is divided into two parts. In the first part a number of electrical and mechanical systems which lead to self-excited oscillations are studied in some detail, and the corresponding differential equations are derived. The remainder of this part is concerned with an analysis of the free oscillations, i.e., those which occur without the action of external exciting forces that depend on the time. This part could thus have been placed in Chapter III, but the importance of the specific problems warrants a separate treatment. The basic occurrence, from the mathematical point of view, is that of limit cycles in the phase plane in the sense of Poincaré. In the simplest case just one such stable limit cycle occurs, and this in turn means that all motions tend to a unique periodic motion. This is in the strongest contrast with the behavior of the free oscillations in systems with nonlinear restoring forces, in which a whole family of free oscillations occurs. If the nonlinearity is small (in other words, if the oscillation is in the neighborhood of the linear oscillation) the problems can be and are treated by the perturbation method. However, the cases in which the departure from linearity is large—even very large—are of particular interest in this class of problems. In such cases the resulting oscillations are of a jerky, not to say discontinuous, character. They are often given the name relaxation oscillations. It is comparatively easy to obtain the lowest order term in an asymptotic development for the period of such a relaxation oscillation in terms of a parameter characterizing the departure from nonlinearity, but it is not easy to obtain higher order terms. Unfortunately it turns out that the higher order terms yield rather large contributions in cases in which the oscillation is markedly of the relaxation type, so that it is important to have a means of calculating them. Such a means has been devised quite recently by Dorodnitsyn, for the van der Pol equation, and more generally by Haag. The details of the complete asymptotic development are carried out in the first part of Chapter V for one relatively simple case.

The second part of Chapter V is concerned with forced oscillations of systems whose free oscillations are of the self-sustained type. It is assumed in all cases that the oscillations do not depart too much from the linear oscillations. This theory was created by van der Pol, who invented a special analytical mode of attacking the problems which is different from any of the several methods treated in Chapter IV. Van der Pol's method is the method used almost exclusively by the Russian writers—particularly Kryloff and Bogoliuboff—for solving all of the various types of problems that involve the time explicitly in the differential equation. The author believes, however, that this method, while it is particularly well suited to deal with the problems treated in this part of Chapter V, is not necessarily the simplest or the most straightforward method of treating other problems. The theory of van der Pol is developed with reference to a particular electric circuit containing a triode vacuum tube and a source of alternating current. In treating the response phenomena, i.e., the amplitude of the periodic response as a function of the frequency and amplitude of the excitation, the elegant variant of van der Pol's method introduced by Andronow and Witt is employed in order to study the stability of all possible periodic oscillations having the frequency of the excitation. This is done by reducing the problem to one of classifying singularities of the first order differential equation of van der Pol in accordance with the criteria of Poincaré derived in Chapter III, and this in turn is made feasible by the fact that to each periodic oscillation there is a corresponding type of singularity. This idea of Andronow and Witt also yields more than the criteria for the stability of the oscillations. By means of it one is led to the possibility of the occurrence of combination oscillations, i.e., oscillations that are the sum of two oscillations, one with the frequency of the excitation, the other with a frequency close to that of the free nonlinear oscillation. Such combination oscillations are correlated with the presence of Poincaré limit cycles of the first order differential equation just mentioned above. The methods used can be extended to prove that the combination oscillations for sufficiently large detuning, i.e., for sufficiently large differences between the frequencies of the free and the forced oscillations, are unique and stable. The circumstances which may occur when the detuning is neither very small nor very large are quite complicated; some description of the phenomena in such cases—which include jump phenomena of various sorts reminiscent of similar phenomena studied in Chapter IV—is given following recent work of Cartwright and Littlewood.

The final Chapter VI of the book returns once more to linear oscillations, but this time the characteristics of the systems treated are assumed to be not constant, as was the case in Chapter I, but rather periodic in the time. As a consequence the differential equations become of the type called Hill's equations. There are several reasons which dictated the inclusion of a lengthy chapter on linear systems in a book devoted primarily to nonlinear systems. In the first place, the treatment of the important question of stability of any periodic nonlinear oscillation leads inevitably to such Hill's equations. Second, the vibration phenomena encountered in systems of this type have features—the occurrence of vibrations somewhat like the subharmonics, for example—which place them, in a sense, in a position between those of nonlinear systems and of linear systems having constant characteristics. A few mechanical and electrical problems leading to Hill's equation are first discussed. Then follows an account of the Floquet theory for linear differential equations with periodic coefficients. For the study of the stability of a given periodic nonlinear oscillation it is necessary to determine whether the solutions of a certain Hill's equation are all bounded or not when certain parameters in the Hill's equation are given. The question of separating the "stable" from the "unstable" parameter values is discussed in detail for the most important special case, the Mathieu equation. These results are then applied to test the stability of the forced oscillations of the Duffing equation, which were treated in Chapter IV, with results the same as were advanced in Chapter IV as the result of plausible physical arguments.

As has been indicated earlier, the Appendices are devoted to a number of mathematical questions which, however, are in some instances also of interest from a practical point of view. Appendix I gives a rigorous treatment of the perturbation method in general as applied to periodic oscillations in the neighborhood of a linear oscillation. Again the basic idea used is due to Poincaré. The general theory is then applied to prove the existence of the perturbation series for all cases treated in this book. In the course of satisfying the conditions needed to ensure the existence of the solutions, one finds that important clues regarding the manner of interpreting the results are uncovered, and that insights are gained regarding the appropriate means of calculating the solutions concretely in the various cases. In Appendix II the existence of combination oscillations of certain systems with nonlinear restoring forces and with viscous damping is proved by obtaining a convergent perturbation series. This result is included as a contrast to the corresponding case in which no damping occurs, when the famous "difficulty of the small divisors" alluded to above makes it practically certain that no convergent perturbation series could be obtained. In Appendix III the existence of a limit cycle in certain rather general systems of the self-sustained type is proved by using a topological method involving the proof of the existence of a fixed point of a certain mapping. This Appendix is modeled somewhat on the work of Levinson and Smith, but a less general case is treated. Appendix IV gives a rigorous proof for the commonly conjectured character of the relaxation oscillation when the departure from nonlinearity becomes very large. Appendix V gives a derivation of the criterion of Poincaré for the stability of a limit cycle, or orbital stability, as it is called. Finally, in Appendix VI, the uniqueness of the limit cycle for the system treated in Appendix III is proved (following the idea of Levinson and Smith) by showing that all possible limit cycles are stable.

Acknowledgments

The writing of this book was greatly facilitated by the generous support of the Office of Naval Research under Contract N6ori-201, Task Order No. 1. The help and encouragement given the author by the ONR in general, and by Dr. Mina Rees in particular, is gratefully acknowledged.

The starting point of this book was a set of lecture notes which were published in 1941 with the aid of a grant from the Rockefeller Foundation. In preparing these lecture notes the author was fortunate in having the help of Professor A. S. Peters.

In the preparation of the present book, the author was greatly aided by Professor E. Isaacson, who corrected many errors, supervised all of the considerable number of calculations, read proofs, and and gave much good advice of all kinds. The drawings were made by Mrs. L. Scheer, and the index by Paul Berg. The author is particularly grateful to Mrs. E. Rodermund, who not only typed the entire manuscript, but also acted as a sharp-eyed critic and detector of innumerable slips of all kinds.

The greater part of what the author has put into this book was learned in the course of seminars conducted in collaboration with his friend and colleague, Professor K. O. Friedrichs. It has been the author's hope that the writing of the book would be a joint enterprise, but this proved impossible. Nevertheless, the author has had the benefit of critical comments by Professor Friedrichs on many parts of the book.

This book is dedicated to R. Courant as a token of esteem and friendship, and as an acknowledgment of the strong influence he has had in the author's scientific development.

New York, N. Y.   J. J. STOKER
January, 1950

Contents

CHAPTER I

Linear Vibrations

1. Introduction

The purpose of this book is the treatment of a variety of non-linear vibration problems. It is of value and interest, however, to summarize briefly the essential features of linear* vibration problems governed by the differential equation

$$m\ddot{x} + c\dot{x} + kx = P(t)$$

in which the coefficients are constants.† There are a number of good reasons for doing so: One of our principal objects is to compare and contrast linear with nonlinear vibration problems. Also, it is the practice to carry over as much as possible of the terminology used in linear problems to the nonlinear problems—a practice which is partially successful but sometimes overdone. Finally, it is useful to have a summary of the chief ideas and formulas of the linear theory available for reference, since the solution of a nonlinear problem is quite often made to depend upon the solution of a sequence of linear problems; this is the case, for example, with the often used perturbation method.

2. Free vibrations

Perhaps the simplest example of a linear vibration problem is furnished by a mechanical system consisting of a mass m attached to a spring which exerts a force (called *the restoring force* or *spring force*)

---

* For complete treatments of linear vibration problems see the books of den Hartog [16] and Timoshenko [38].

† Here and throughout this book dots over a quantity refer to differentiations with respect to the time.

1

proportional to the displacement x of the mass (see Figure 2.1). If, in addition, the mass is considered to move in a medium which exerts

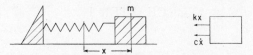

FIG. 2.1. Linear mechanical system.

a resistance proportional to the velocity (*a viscous damping force*), the equation of motion is

(2.1) $$m\ddot{x} + c\dot{x} + kx = 0$$

in which m, c, and k are positive constants. If we introduce the quantities

(2.2) $$r = c/2m, \qquad p^2 = k/m, \qquad q = \sqrt{p^2 - r^2},$$

(2.1) becomes

(2.3) $$\ddot{x} + 2r\dot{x} + p^2 x = 0.$$

The general solution of this linear homogeneous differential equation with constant coefficients is a linear combination of two exponential functions, as follows

(2.4) $$x = Ae^{\lambda_1 t} + Be^{\lambda_2 t}$$

in which A and B are arbitrary constants and λ_1 and λ_2 are roots of the quadratic

(2.5) $$\lambda^2 + 2r\lambda + p^2 = 0.$$

Hence $\lambda_{1,2}$ are given by

(2.6) $$\lambda_{1,2} = -r \pm iq, \qquad i = \sqrt{-1}.$$

Since we wish to express the solution (2.4) in real form, we consider the three cases in which q is: (a) real, (b) zero, (c) imaginary. The solutions in real form corresponding to these cases are readily found to be

(2.7) $$\begin{cases} \text{a) } x = e^{-rt}(c_1 \cos qt + c_2 \sin qt), \\ \text{b) } x = e^{-rt}(c_1 t + c_2), \\ \text{c) } x = c_1 e^{\lambda_1 t} + c_2 e^{\lambda_2 t}, \qquad \lambda_1, \lambda_2 \text{ real.} \end{cases}$$

The arbitrary constants are now c_1 and c_2; in the applications their values are normally fixed by prescribing the values of the displacement x and the velocity $\dot{x}$ at some time $t = t_0$.

Case (a) of (2.7) occurs most frequently in the applications. As one sees from (2.2), this case occurs if r (the *damping constant*) is small compared with p^2. The equation (2.7a) then represents an oscillatory motion such that any two successive maxima x_1 amd x_2 of the displacement satisfy the relation

$$x_2 = e^{-2\pi r/q} x_1 .$$

Thus if $r > 0$, the motion dies out exponentially (the time t being always taken positive, of course), but if $r < 0$ (in which case the damping is sometimes said to be negative), the oscillations increase exponentially. The commonest cases are of course those in which $r \geq 0$.

If $r = 0$ the system is said to be without damping. For this important special case the resulting motion is given by

$$(2.8) \qquad\qquad x = c_1 \cos pt + c_2 \sin pt.$$

Thus a simple harmonic motion results in which the *circular frequency* p is determined by $p = \sqrt{k/m}$. For q real, the oscillations given by (2.2) and (2.4) are called *free* or *natural* oscillations. The quantity $p/2\pi$ is called the *natural frequency*.

The solution (2.7b) corresponding to $q = 0$ corresponds to the transition from the motions of oscillatory character given by (2.7a) to the motions given by (2.7c) which are not oscillatory; the motion is said to be one corresponding to *critical damping*. The nonoscillatory motions occur, of course, when the damping coefficient is relatively large.

3. Forced vibrations

Consider now the motion which results when an external force $P(t)$ depending only on the time is applied to the system of Section 2, in addition to the other forces. The equation of motion is then the nonhomogeneous linear differential equation

$$(3.1) \qquad\qquad m\ddot{x} + c\dot{x} + kx = P(t).$$

The most important case for our purposes is that in which $P(t)$ is periodic. It will be assumed here that $P(t)$ is a simple harmonic function given by

$$(3.2) \qquad P(t) = F \cos (\omega t + \varphi)$$

in which F is the *amplitude*, ω the *circular frequency*,* and φ a constant called the *phase* of $P(t)$. The solutions of (3.1) consist of the sum of the solutions of the homogeneous equation (i.e. the free oscillations discussed in the preceding section) and of any solution of the non-homogeneous equation. If we assume that the free oscillation is of the type given by (2.7a), the solutions of (3.1) with $P(t)$ given by (3.2) are readily obtained in the form

$$(3.3) \qquad x = e^{-rt}(c_1 \cos qt + c_2 \sin qt) + \frac{F \cos (\omega t + \varphi - \delta)}{m\sqrt{(p^2 - \omega^2)^2 + 4r^2\omega^2}}.$$

The square root in the denominator is zero only if $p = \omega$ and $r = 0$, and this constitutes an important special case to be treated later. In other words, the resulting motion is obtained by a superposition of the free oscillation and an oscillation called the *forced oscillation* which arises from the action of the external force.

We observe that *the frequency of the forced oscillation is the same as that of the external force.* The amplitude H of the forced oscillation is given by

$$(3.4) \qquad H = F/m\sqrt{(p^2 - \omega^2)^2 + 4r^2\omega^2},$$

while the phase δ—or, better, the phase shift relative to the phase of the external force $P(t)$—is given by

$$(3.5) \qquad \begin{cases} \cos \delta = (p^2 - \omega^2)/\sqrt{(p^2 - \omega^2)^2 + 4r^2\omega^2}, \\ \sin \delta = -2r\omega/\sqrt{(p^2 - \omega^2)^2 + 4r^2\omega^2}. \end{cases}$$

In the case of positive damping ($r > 0$), it is clear from (3.3) that after a sufficiently long period of time *the free oscillation is damped out and only the forced oscillation would be observed.*

In the present connection it is important to consider the special case $r = 0$ in which there is no damping. In this case the phase shift δ is seen from (3.5) to be zero for $\omega < p$ and π for $\omega > p$; in other words the forced oscillation is in phase with the external force

---

* For the sake of brevity we shall often refer, both here and later on, to ω as the frequency rather than the circular frequency if no confusion is likely to arise.

if the free or natural frequency p is greater than the frequency of the external force and is 180° out of phase with it when $\omega > p$. For $r = 0$, we obtain from (3.3) the oscillation

$$(3.6) \qquad x = (c_1 \cos pt + c_2 \cos pt) + \frac{F}{m \mid p^2 - \omega^2 \mid} \cos (\omega t + \varphi - \delta)$$

provided that $p \neq \omega$. The resulting oscillation is the superposition of two simple harmonic motions, one with the natural frequency and the other with the frequency of the external force. The case $p = \omega$ in which the free and the forced oscillations have the same frequency, the case of *resonance*, is of great interest and importance; the solution of (3.1) for $r = 0$, $p = \omega$ is found to be

$$(3.7) \qquad x = c_1 \cos \omega t + c_2 \sin \omega t - \frac{Ft}{2\omega m} \sin (\omega t + \varphi).$$

We observe that the motion due to the external force is no longer periodic, but is oscillatory with an "amplitude" $Ft/2\omega m$ which increases linearly with the time. One of the principal reasons for which vibration phenomena are important for the applications lies in the possibility of the occurrence of such resonance phenomena: It is often vital to design machine parts or other engineering structures in such a way as to avoid resonance with periodic forces which may be impressed on the system (through slight unbalances in rotating machinery, for example) because of the danger that the structure might be destroyed through the building up of vibrations with large amplitudes.

If there is damping in the system it is clear from (3.3) that the amplitude of the motion is always finite (for $r > 0$, of course). It is nevertheless of interest in practice to investigate the amplitude of the forced oscillation since a rupture of the spring would occur if the amplitude were to become too large. The numerical value $\mid H \mid$ of the amplitude of the forced oscillation can be written in the form $\mid H \mid = M(F/k)$ in terms of the *magnification factor* M and the static deflection (F/k) of the system under a constant force F equal to the amplitude of the given periodic external force. From (3.4) and $p^2 = k/m$ we find readily for the magnification factor the relation

$$(3.8) \qquad M = \frac{1}{\sqrt{\left(1 - \dfrac{\omega^2}{p^2}\right)^2 + \left(2\,\dfrac{r}{p}\right)^2 \dfrac{\omega^2}{p^2}}}.$$

The extreme values for M are attained for $\omega = 0$ and $\omega^2/p^2 = 1 - 2(r/p)^2$. If $1 - 2(r/p)^2 < 0$, there is a maximum for $\omega = 0$;

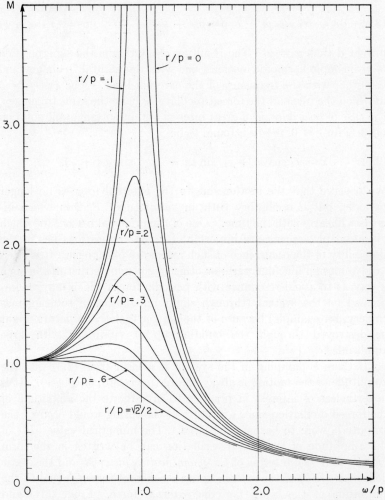

FIG. 3.1. Response curves for the linear forced oscillation.

if $1 - 2(r/p)^2 > 0$ and $\omega > 0$, there is a maximum for $\omega/p = \sqrt{1 - 2(r/p)^2}$ and a minimum for $\omega = 0$. For small values of the damping constant r, the frequency which produces the maximum amplitude is very nearly the natural frequency. Figure 3.1 shows curves for M as a function of ω/p for various values of r/p. We shall refer to these curves as *response curves*, since they obviously yield the amplitude, or response, of the system for an external force of any given frequency.

4. Subharmonics and ultraharmonics

If there is damping in the linear system treated above, we see from (3.3) that no periodic motion exists unless the initial conditions are chosen in such a way that the damped free oscillation does not occur,* and in this case the motion has the same period as the external force. Even if the free oscillation does occur, it is always damped out in actual physical cases, so that the eventual steady motion has the same period as the applied force. In nonlinear systems, however, the circumstances may be quite different in this respect: We shall see later that nonlinear systems can possess a wide variety of periodic motions in addition to those which have the same period as the external force; for example, what are called subharmonic oscillations can occur in which the smallest period of the motion may be any integral multiple of the period of the external force.

If we assume our linear system to be undamped, it is possible under special circumstances to obtain all of the various types of periodic motions which are encountered in nonlinear systems†; it therefore seems worth while to classify here the possibilities in this respect and to introduce the terminology to be used later. We consider the differential equation

$$(4.1) \qquad \ddot{x} + p^2 x = F \cos \omega t.$$

Since we are interested only in periodic solutions, we may without loss of generality choose the time $t = 0$ such that $\dot{x}(0) = 0$. If $x(0) = A$, the solution of (4.1) is

$$(4.2) \qquad x(t) = \left(A + \frac{F}{\omega^2 - p^2} \right) \cos pt - \frac{F}{\omega^2 - p^2} \cos \omega t.$$

---

* It is easy to see that the initial conditions can always be chosen to accomplish this.

† This remark is the starting point of an investigation by M. Levenson [23] in which the main purpose is a study of the periodic solutions of a nonlinear differential equation.

The solutions $x(t)$ are periodic in the following cases (and only in these cases):

(a) $B = A + \dfrac{F}{\omega^2 - p^2} = 0$.   The free oscillation does not occur.

(b) $\omega = np$, n any integer except $n = 1$, $B \neq 0$.

(c) $\omega = p/m$, m any integer except $m = 1$, $B \neq 0$.

(d) $\omega = np/m$, n, m relatively prime integers, $B \neq 0$.

In case (a) the solution $x(t)$ has obviously the period $2\pi/\omega$ of the external force, which is quite arbitrary. However, the amplitude of this oscillation depends on ω. We call such an oscillation a *harmonic oscillation*.

In case (b) the solution $x(t)$ has as its least period the period $2\pi/p$ of the free oscillation, which is n times the least period $2\pi/np$ of the external force. Such an oscillation is called a *subharmonic oscillation of order n*, or simply a *subharmonic*.

In case (c) the solution $x(t)$ has the same period $2m\pi/p$ as the external force, just as in case (a) of the harmonic solution, but the two cases are different since the present case is a superposition of two solutions whose least periods differ, while the harmonic case, as its name implies, consists of a single harmonic. The oscillations in case (c) will be referred to as *ultraharmonics*.

In case (d) the solution $x(t)$ has the period $2\pi m/p$ while the external force has the period $2\pi m/np$; thus the least period of the oscillation is n times that of the external force, just as in case (b) of the *subharmonics*. However, the period of the oscillation in the present case is not the same as that of the free oscillation but is rather m times it. For lack of a better term we call such oscillations *ultra-subharmonics*.

Since any actual system always involves some damping, it is clear that cases (b), (c), and (d) above, which require the coexistence of the free and the forced oscillations, would never be observed in practice. In nonlinear systems, however, it is a matter of great theoretical as well as practical significance that all four cases may occur even if viscous damping is present.

In Figure 4.1 the four types of periodic oscillations discussed above are indicated schematically.

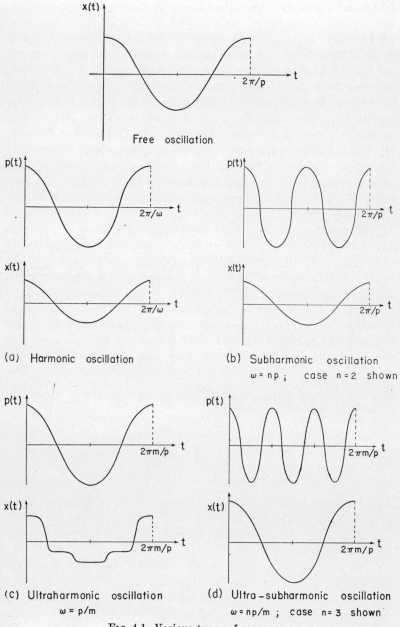

Fig. 4.1. Various types of response.

5. Linear systems with variable coefficients

The linear system considered so far was one with constant coefficients. There are, however, important cases in which one or more of the quantities determining the nature of the system, i.e. the mass, damping factor, and spring stiffness, may depend upon the time; in such cases the differential equation remains linear but has variable coefficients. The vibratory motions which occur can still be said to be the result of a superposition of a "free" and of a "forced" oscillation. However, the theory of linear systems with nonconstant coefficients is quite different and much more complicated than that of the systems with constant coefficients. For example, the subharmonics and ultraharmonics of all orders may occur even if viscous damping is present.

Perhaps the most important special case of a linear system with variable coefficients is that in which the coefficients vary periodically in the time. For this case a well-rounded and rather complete theory exists which, aside from its intrinsic interest, also is of great importance for the discussion of the stability of the periodic solutions of nonlinear systems. We therefore postpone the discussion of the theory of linear differential equations with periodic coefficients until Chapter VI, where it is treated in considerable detail and applied to certain stability problems arising from nonlinear vibrating systems.

6. Principle of superposition for linear systems. Contrast with nonlinear systems

If x_1 is a solution of $m\ddot{x} + c\dot{x} + kx = P_1(t)$ and x_2 is a solution of $m\ddot{x} + c\dot{x} + kx = P_2(t)$, then $x = x_1 + x_2$ is a solution of

$$m\ddot{x} + c\dot{x} + kx = P_1(t) + P_2(t).$$

This fundamental fact, a direct consequence of the linearity of the differential equation, is called the *principle of superposition*. It is important to note explicitly that the principle does not hold for nonlinear differential equations.

The notions of free and forced oscillation, free and forced frequency, and resonance, are intimately related to the principle of superposition and thus have real sense only for linear systems. It has become more or less standard practice to refer to these terms in the analysis of vibration problems which lead to nonlinear differential equations. There are occasions when the use of these terms is helpful in nonlinear problems; for example, Figure 3.1 has its counterpart in certain nonlinear cases. One should, however, exercise some caution in the use of these terms borrowed from linear vibration theory.

The theory of linear differential equations has been very thoroughly studied and developed. As a consequence the theory of linear vibrations is, in a sense, a closed and well-rounded one, particularly for systems with constant coefficients. On the other hand, there is almost nothing of such a general character known about nonlinear differential equations. The analysis of nonlinear vibration problems therefore depends largely upon the use of approximation methods, and it is confined for the most part to the discussion of a variety of special cases.

CHAPTER II

Free Vibrations of Undamped Systems with Nonlinear Restoring Forces

1. Classification of Problems

The main purpose of this book is the treatment of mechanical or electrical systems which are governed by the differential equation

$$(1.1) \qquad m\ddot{x} + \varphi(\dot{x}) + f(x) = F \cos \omega t.$$

In analogy with the linear system treated in the first chapter we shall frequently refer to the term $m\ddot{x}$ as the *inertia force*, to $-\varphi(\dot{x})$ as the *damping force*, to $-f(x)$ as the *restoring force* or *spring force*, and to the term $F \cos \omega t$ (with F and ω constant) as the *external force* or *excitation*. We shall have many occasions later to see how the terms in (1.1) can be interpreted in the case of electrical and combined electrical-mechanical systems of various types.

We do not study equation (1.1) in all generality, i.e. for arbitrary nonlinear damping forces $-\varphi(\dot{x})$ and restoring forces $-f(x)$. Such knowledge as we have about the vibration phenomena associated with (1.1) is largely confined to certain special cases, each one of which is treated in a separate chapter of this book.

We proceed to classify the special cases treated in the remainder of this book:

$$(a) \qquad \ddot{x} + f(x) = 0.$$

We refer to the motions which result in this case as undamped (since $\varphi(\dot{x}) \equiv 0$) and free (since $F \cos \omega t \equiv 0$) vibrations of a system with a nonlinear restoring force $-f(x)$. This simplest of our cases is treated in the present chapter.

$$(b) \qquad \ddot{x} + \varphi(\dot{x}) + f(x) = 0, \quad \dot{x}\varphi(\dot{x}) \geq 0.$$

13

The motions in this case are of course referred to as free vibrations with damping. The condition $\dot{x}\varphi(\dot{x}) \geq 0$ on the function φ ensures that the damping force acts in a direction opposite to the velocity, and hence is a true resistance. This case, which comprises a wide variety of interesting problems, is treated in Chapter III. At the same time, a few problems which might have been treated already in Chapter II are treated here in order to illustrate the methods.

(c) $$\ddot{x} + c\dot{x} + f(x) = F \cos \omega t, \qquad c \geq 0.$$

Here the motions are quite naturally referred to as forced oscillations of systems with a nonlinear restoring force. We also refer to the equation on occasion as the Duffing equation, since Duffing [9] made the first significant progress in studying it. The problems concerned with this equation are treated in Chapter IV.

(d) $$\ddot{x} + \varphi(\dot{x}) + x = F \cos \omega t, \qquad \begin{array}{l} \dot{x}\varphi(\dot{x}) < 0 \text{ for } \dot{x} \text{ small,} \\ \dot{x}\varphi(\dot{x}) > 0 \text{ for } \dot{x} \text{ large.} \end{array}$$

The motions in this case are referred to as self-excited oscillations— free oscillations if $F \cos \omega t \equiv 0$ and forced oscillations otherwise. The reason for the adjective *self-excited* is that the "damping" force is in the direction of the velocity for small velocities so that the state of rest is not stable and a motion will develop from the rest position under the slightest disturbance even if the excitation $F \cos \omega t$ is zero. On the other hand, the free oscillations are limited in amplitude eventually because of the fact that the damping force is assumed to be positive, i.e. to be opposite in direction to the velocity, when the velocity is above a certain value. The differential equation in the present case (or, rather, the special case of it in which $\varphi(\dot{x}) = -\dot{x} + \frac{1}{3}\dot{x}^3$) is commonly referred to as Rayleigh's equation or the van der Pol equation (cf. [32]). We study this case in Chapter V.

2. Examples of systems governed by $\ddot{x} + f(x) = 0$

The best known example of a vibratory motion which is governed by such an equation is that of the simple pendulum. Its motion is determined by

(2.1) $$ml^2\ddot{x} + mgl \sin x = 0$$

where l is the length of the pendulum, m is the attached mass, x is the angular displacement, and g is the acceleration of gravity. If it is assumed that x is so small that $\sin x$ can be replaced by x with sufficient accuracy, (2.1) can be replaced by

$$(2.2) \qquad ml^2\ddot{x} + mglx = 0.$$

As we know from the preceding chapter, this implies that the motion is isochronous with period $T = 2\pi\sqrt{l/g}$, that is, the period is independent of the initial velocity and displacement. For large displacements (2.2) is of course inaccurate; however, for displacements ranging up to about one radian, $\sin x$ in (2.1) can be replaced with fair accuracy

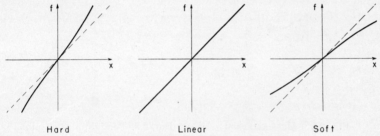

Hard   Linear   Soft

FIG. 2.1. The three types of spring forces.

by the first two terms in its Maclaurin's expansion. The equation then becomes

$$(2.3) \qquad ml^2\ddot{x} + mgl\left(x - \frac{x^3}{6}\right) = 0.$$

It will be seen later that the motions defined by this equation and (2.1) are not isochronous.

The quantity $-f(x)$ is the force exerted by the spring when it is subjected to a displacement x. The *stiffness* of the spring for a displacement x may be defined as $f'(x)$. If the stiffness increases with the displacement, the spring is said to be *hard*. If the stiffness decreases as the displacement increases, the spring is called a *soft* spring. In equation (2.3) the force $mgl(x - x^3/6)$ is obviously one with decreasing stiffness, i.e., it is a soft spring force. Figure 2.1 indicates the distinction between types of restoring forces.

Another example of a system with a nonlinear restoring force is provided by a mass m which is attached to the middle point of a stretched wire as shown in Figure 2.2. For a horizontal displacement x of the mass the strain ϵ in the wire is $\epsilon = (\sqrt{a^2 + x^2} - a)/a$. If S is the initial tension in the straight wire, the tension in it after the displacement x is $S + AE\epsilon$, in which E is the modulus of elasticity of the wire and A is its cross section area. The restoring force on the mass m in the x-direction is $2(S + AE\epsilon) \sin \varphi$, with $\sin \varphi = x/\sqrt{a^2 + x^2}$. We assume now that x is so small compared with a that we may develop the restoring force in powers of x/a and neglect

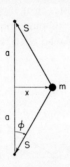

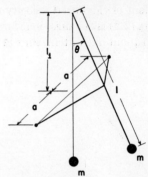

FIG. 2.2. Mass attached to a stretched wire.   FIG. 2.3. Pendulum constrained by a stretched wire.

all terms of order higher than the third in x/a; the differential equation governing the motion is readily found to be

$$(2.4) \qquad m\ddot{x} + \frac{2S}{a} x + \frac{AE}{a^3} \left(1 - \frac{S}{AE}\right) x^3 = 0.$$

Since the coefficient of x^3 is positive it is clear that the restoring force is hard in this case—it would be legitimate in most cases to neglect the term S/AE in the coefficient of the x^3 term since it represents the initial strain in the wire and hence would be small compared with unity.

It is interesting to note that the approximate differential equation which determines the motion of a pendulum attached to a stretched wire perpendicular to the plane of motion, as shown in Figure 2.3, is found, after a little calculation, to be

$$(2.5) \qquad ml^2\ddot{\theta} + \left(mgl + \frac{2Sl_1^2}{a}\right) \theta$$
$$- \left(\frac{mgl}{6} + \frac{Sl_1^2}{3a} + \frac{Sl_1^4}{a^3} - \frac{EAl_1^4}{a^3}\right) \theta^3 = 0.$$

The quantities S, E, and A have the same meaning as in the preceding example. By choosing the parameters properly, for example by taking for EA the value

$$(2.6) \qquad EA = S + \frac{Sa^2}{3l_1^2} + \frac{mgla^3}{6l_1^4},$$

it is possible to make the coefficient of θ^3 vanish. In this case the combination of stretched wire and pendulum would afford a good approximation to a system having isochronous oscillations.

As an example of a nonlinear electrical problem, take the problem of finding the flux variation in an iron core inductance coil which is connected to a charged condenser as shown in Figure 2.4. If φ is the

FIG. 2.4. Circuit with capacitance and nonlinear inductance.

flux linking the coil the equation of the circuit may be written in the form

$$\frac{d\varphi}{dt} + \frac{q}{c} = 0$$

where t is the time, q is the charge on the condenser, and c is its capacitance. Since the current i is equal to dq/dt, the above equation takes the form

$$(2.7) \qquad \frac{d^2\varphi}{dt^2} + \frac{i}{c} = 0.$$

In elementary circuit theory the linear relation $i = \varphi/L$ between current and flux is assumed, in which L is the inductance. A more accurate relation between the current and flux in an iron core inductance is given by $i = \alpha\varphi - \beta\varphi^3$; $\alpha > 0, \beta > 0$. If this function is substituted for i in (2.7) we obtain the nonlinear differential equation

$$\frac{d^2\varphi}{dt^2} + \frac{1}{c}(\alpha\varphi - \beta\varphi^3) = 0,$$

which is of the type we consider in this chapter.

In Sections 8, 9, and 10 of the next chapter still other examples of problems governed by the equation $\ddot{x} + f(x) = 0$ are given.

3. Integration of the equation $m\ddot{x} + f(x) = 0$

A first integral of $m\ddot{x} + f(x) = 0$ can be easily obtained, since the substitution

$$v = \frac{dx}{dt} \; ; \; \frac{d^2 x}{dt^2} = \frac{dv}{dx} \cdot \frac{dx}{dt} = v \frac{dv}{dx}$$

reduces this equation to the first order differential equation

$$(3.1) \qquad mv \frac{dv}{dx} + f(x) = 0$$

in which the variables are separable.   Thus

$$mv\,dv = -f(x)dx$$

and if $v = v_0$ when $x = x_0$, the velocity v is given by the equation

$$(3.2) \qquad \frac{mv^2}{2} - \frac{mv_0^2}{2} = -\int_{x_0}^{x} f(x)\,dx = -\,[F(x) - F(x_0)]$$

which expresses the law of conservation of energy.   The left side of the equation represents the change in kinetic energy; the right side represents the work done by the restoring force, or the change in potential energy.   From (3.2) we find for the velocity v the equation

$$(3.3) \qquad v = \frac{dx}{dt} = \sqrt{v_0^2 + \frac{2}{m}\left(F(x_0) - F(x)\right)}.$$

The sign of the square root must of course be appropriately chosen. The time t can be found as a function of the displacement by separating the variables again and integrating.   Hence if the initial value of t is t_0 we have

$$(3.4) \qquad t = \int_{x_0}^{x} \frac{dx}{\sqrt{v_0^2 + \dfrac{2}{m}\,[F(x_0) - F(x)]}} + t_0 \,.$$

However, it is understood here that it is in general necessary to pass from one branch of the square root to the other whenever $v = dx/dt$ passes through the value zero.   The curves given by (3.2) in the x, v-plane are curves of constant energy; we refer to them frequently as *energy curves*.

4. Geometrical discussion of the energy curves in the phase plane

In this section we discuss the curves of constant energy in the x, v-plane, which we call the *phase plane*. As we shall see, important information about the motion of a qualitative character can be obtained rather easily. We note that x and $v = dx/dt$ are functions of t, that is, the curves in the x,v-plane may be regarded as given in parametric form with t as parameter. From $v = dx/dt$ it follows, then, that x increases with t when v is positive.

The closed energy curves (we refer to the energy curves also as integral curves) are particularly important since they correspond to periodic motions $x(t)$: If $x(t)$ is periodic it is obvious that the corresponding x,v-curve is closed. On the other hand, if an x,v-curve is closed, it follows that the displacement and velocity at any time t are reached again after a certain time T, i.e., $x(t + T) = x(t)$ and $v(t + T) = v(t)$ and the motion is evidently periodic. The period T can obviously be calculated by the line integral

$$T = \oint \frac{dx}{v}$$

to be taken along the closed integral curve in the direction of increasing t.

Consider first the case in which $f(x)$ is linear. With $f(x) = kx$ the differential equation for the x,v-curves is

$$v \frac{dv}{dx} = -kx$$

from which we obtain the solution

$$(4.1) \qquad v^2 + kx^2 = v_0^2 + kx_0^2$$

in which x_0 and v_0 are the initial displacement and velocity. All integral curves are ellipses if $k > 0$, which we now assume, and hence every motion is periodic. The motion is, as we know from the preceding chapter, a simple harmonic motion and $x(t)$ and $v(t)$ are given by

$$(4.2) \qquad \begin{cases} x = a \cos pt \\ v = -ap \sin pt \end{cases}$$

with $p^2 = k$. The amplitude a (that is, the maximum displacement) should evidently be given by $a = \sqrt{(v_0^2 + kx_0^2)/k}$ in view of (4.1). Equation (4.2) can, of course, be obtained by setting $v = dx/dt$ in (4.1) and integrating once more under the conditions that $x = a$ and $v = 0$ for $t = 0$. The period T of the motion is $2\pi/p$; it could also be obtained from

$$T = \oint \frac{dx}{v} = \frac{4}{p} \int_0^a \frac{dx}{\sqrt{a^2 - x^2}} = \frac{4}{p} \int_0^1 \frac{dy}{\sqrt{1 - y^2}} = \frac{2\pi}{p}.$$

In this linear case we note the well known fact that the period of the motion is independent of the amplitude, or as we could also put it, *all of the closed solution curves in the phase plane are traversed in the same time.* If we had taken for k the negative sign in the above discussion, the curves (4.1) would be hyperbolas and no periodic motions would exist.

We consider next the more general case of a spring force $-f(x)$ of the form

$$(4.3) \qquad\qquad f(x) = \alpha x + \beta x^3, \qquad\qquad \alpha > 0.$$

In this nonlinear case the determination of x and v as functions of t is still possible by explicit integration through the use of elliptic integrals and functions. However, it is quite possible to discuss the qualitative nature of the motions $x(t)$ which result from the differential equation

$$(4.4) \qquad\qquad v \frac{dv}{dx} = -(\alpha x + \beta x^3)$$

rather easily without explicit use of elliptic integrals. An integration yields the equation

$$(4.5) \qquad\qquad v^2 + \alpha x^2 + \beta \frac{x^4}{2} = h = \text{constant},$$

in which the constant h represents twice the total energy in the system. In the neighborhood of the origin $x = 0$, $v = 0$ the curves given by (4.5) are all closed curves which have the appearance of ellipses, since $\beta x^4/2$ can be neglected in comparison with αx^2 for small x; the constant h is then of necessity small and positive. The maximum displacement $x_{\max} = a$ is readily seen to satisfy the relation

$$(4.6) \qquad\qquad a^2 = \frac{-\alpha + \sqrt{\alpha^2 + 2\beta h}}{\beta}, \qquad \diagup$$

obtained by setting $v = 0$, in which the positive sign of the radical is taken for $\beta > 0$ (hard spring) as well as for $\beta < 0$ (soft spring) since $h > 0$ and a^2 should be small and positive. On account of the symmetry of the closed phase curves (4.5) the period T of the motion can be obtained (cf. (3.4)) in the form

$$(4.7) \qquad T = 4 \int_0^a \frac{dx}{\sqrt{h - (\alpha x^2 + \beta x^4/2)}} .$$

It is useful to transform the integral in this equation by changing the variable of integration. Since a^2 is a root of $h - (\alpha z + \beta z^2/2) = 0$ we may write

$$(4.8) \qquad h - (\alpha x^2 + \beta x^4/2) = \frac{\beta}{2} (a^2 - x^2)(b^2 + x^2)$$

in which

$$(4.9) \qquad \frac{\beta}{2} (-b^2 + a^2) = -\alpha, \qquad \text{or} \qquad \beta b^2 = \beta a^2 + 2\alpha.$$

If we now introduce θ as new integration variable replacing x in (4.7) by the relation

$$(4.10) \qquad\qquad\qquad x = a \sin \theta,$$

the integral is readily found to take the form

$$(4.11) \qquad T = 4\sqrt{2} \int_0^{\pi/2} \frac{d\theta}{\sqrt{2\alpha + \beta a^2 + \beta a^2 \sin^2 \theta}} ,$$

upon making use of (4.9) to eliminate b^2. From (4.11) we now see at once that the period T of the oscillation in the present cases, unlike the case of the linear oscillations obtained for $\beta = 0$, is not independent of the amplitude a of the oscillation. In fact, for the hard spring ($\beta > 0$) the period is seen to decrease and thus the frequency to increase when the amplitude increases, while just the opposite effect of increasing period and decreasing frequency occurs with increase of the amplitude when the spring is soft ($\beta < 0$). For $\beta > 0$, the phase curves given by (4.5) are all closed curves, in other words all motions are periodic in the case of the hard spring; in this case (4.11) is valid under all circumstances. If $\beta < 0$, however, the curves (4.5) are closed curves only in a certain region of the x,v-plane containing the origin in its interior, so that (4.11) has a meaning in this case only if the quantity a given by (4.6) is not too large.

In Figure 4.1 we show curves indicating schematically the relation between the amplitude a and the circular frequency $\omega = 2\pi/T$ in the three cases of linear, hard, and soft springs. These curves have a common tangent at $\omega = \sqrt{\alpha}$ (the frequency when $\beta = 0$), as one could easily prove using (4.11). This behavior of the amplitude of the free oscillations as a function of frequency is so important that we emphasize once more that *the amplitude increases with the frequency with a hard spring, is independent of the frequency with a linear spring, and decreases with the frequency in the case of a soft spring.*

It is useful and instructive to discuss the curves furnished by (4.5) in more detail, particularly with a view to comparison of the difference between the effects of hard and soft springs. Sets of curves

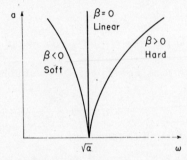

FIG. 4.1. Dependence of amplitude on frequency with different types of spring forces.

for the two cases are shown in Figure 4.2. In the case of the hard spring, little remains to be said since all of the curves are closed curves. The arrows on the curves indicate the direction in which any point $(x(t), v(t))$ moves with increasing t; these directions are obtained at once from $v = dx/dt$, so that, for example, x increases with t when v is positive. We observe also that the curves cross the coordinate axes orthogonally except at points on the x-axis where $f(x) = 0$. For a soft spring the circumstances are more complicated. In this case we may write equation (4.5) in the form

$$(4.12) \qquad v^2 + \alpha x^2 - \rho^2 \frac{x^4}{2} = h, \qquad \rho^2 = -\beta.$$

If x is near to zero we have already noted that $h \geq 0$ and the curves

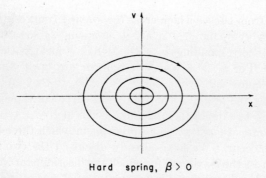

Hard   spring, $\beta > 0$

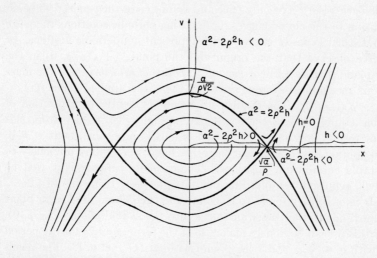

Soft   spring, $\beta < 0$

F$_{\text{I}G}$. 4.2. Velocity-displacement plane for hard and soft spring forces.

given by (4.12) are approximately ellipses. If we solve (4.12) for v^2 we have

$$(4.13) \qquad v^2 = h - \alpha x^2 + \rho^2 \frac{x^4}{2}.$$

If $h > 0$, the curves given by (4.13) cross the v-axis, $x = 0$, with $v = \pm v_0$, $v_0 > 0$. The right-hand side of equation (4.13) is quadratic in x^2 and has $\alpha^2 - 2\rho^2 h$ as discriminant; it is therefore always positive if $\alpha^2 - 2\rho^2 h$ is negative and in this case h must of necessity be positive.

The transition from curves which cross the x-axis to curves which do not cross it thus occurs for $\alpha^2 - 2\rho^2 h = 0$ or $h = \alpha^2/2\rho^2$; on this curve the velocity v_0 corresponding to $x = 0$ has evidently the value $v_0 = \alpha/\rho\sqrt{2}$. For $v_0 > \alpha/\rho\sqrt{2}$ the curves are open curves, as indicated in Figure 4.2, while for $v_0 < \alpha/\rho\sqrt{2}$ they are closed curves encircling the origin. The transition curve corresponding to $v_0 = \alpha/\rho\sqrt{2}$ has two double points on the x-axis at $x = \pm\sqrt{\alpha}/\rho$, and this curve separates the x,v-plane into regions in which three distinct types of curves occur. We have already discussed the two sorts of curves which cross the v-axis and for which the discriminant $\alpha^2 - 2\rho^2 h$ is negative. If the discriminant is positive, a set of curves is obtained which cross the x-axis only once and do not cross the v-axis. Thus, in addition to the closed curves representing periodic solutions there are two distinct types of curves which represent nonperiodic motions.

We observe that the origin and the points $(\pm\sqrt{\alpha}/\rho,\ 0)$ in the x, v-plane (Figure 4.2) correspond to *points of equilibrium* of the mass in our mechanical system since for them the applied force $f(x) = \alpha x - \rho^2 x^3 = 0$. The origin corresponds evidently to a position of stable equilibrium in the sense that a slight disturbance from this point results only in an oscillation of small amplitude about $x = 0$. The points $(\pm\sqrt{\alpha}/\rho, 0)$, on the other hand, correspond obviously to unstable equilibrium positions. These *saddle points* might well be called "repulsive" equilibrium points, as one sees from Figure 4.2.

It is of some interest to calculate the time T required for a point to move along the transition curve into the saddle point $(\sqrt{\alpha}/\rho, 0)$. For this we can obviously make use of the formula (4.7) for the period T, with $a = \sqrt{\alpha}/\rho$, $\beta = -\rho^2$, and without the factor 4. The result is

$$T_S = \frac{\sqrt{2}}{\sqrt{\alpha}} \int_0^{\pi/2} \frac{d\theta}{\sqrt{1 - \sin^2\theta}} = \sqrt{\frac{2}{\alpha}} \int_0^{\pi/2} \sec\theta\, d\theta.$$

We see that T_S is infinite, that is, the time required to attain the unstable equilibrium position is infinite. It is then clear also that any motion in the neighborhood of such an equilibrium point is a slow "creeping" motion. This is a result that holds good in general for equilibrium points of this character.

Finally we consider the case of the pendulum, for which $f(x) = \alpha \sin x$, $\alpha > 0$. The differential equation for the x,v-curves is

$$v \frac{dv}{dx} = -\alpha \sin x$$

and the curves themselves are given by

(4.14) $$v^2 = 2\alpha \cos x + h.$$

These curves are shown in Figure 4.3. We observe that $h > -2\alpha$ must be required or v^2 will be always negative. If $-2\alpha < h < 2\alpha$ one sees readily that the curves (4.14) are closed curves encircling the points $v = 0$, $x = 2n\pi$ (n any integer, positive or negative); in these cases the amplitude a is then seen to satisfy the relation

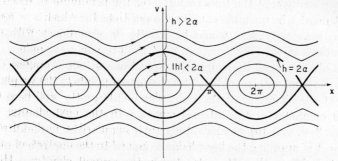

FIG. 4.3. Velocity-displacement plane for the simple pendulum.

$0 = 2\alpha \cos a + h$ and the period T of the oscillations which they represent is given by

$$T = 4 \int_0^a \frac{dx}{\sqrt{2\alpha \, \cos x + h}} = \frac{4}{\sqrt{2\alpha}} \int_0^a \frac{dx}{\sqrt{\cos x - \cos a}}.$$

If a new integration variable replacing x is introduced by the relation $\sin x/2 = \sin a/2 \sin \theta$, the integral for T takes the form

(4.15) $$T = \frac{4}{\sqrt{\alpha}} \int_0^{\pi/2} \frac{d\theta}{\sqrt{1 - \sin^2 \dfrac{a}{2} \sin^2 \theta}},$$

that is, T is given in terms of the complete elliptic integral of the first kind. We observe that the period increases with the amplitude, in accordance with the fact that the spring force is soft in this case.

If $h > 2\alpha$, we observe from (4.14) that v never becomes zero, the curves are open curves as indicated in Figure 4.3. The transition from closed to open curves occurs for $h = 2\alpha$, and v is then given by $v^2 = 4\alpha \cos^2 x/2$. This curve is drawn more heavily than the others in the figure. The situation is similar to that of the soft spring in the preceding example (cf. Figure 4.2), except that there are an infinite number of equilibrium positions which are alternately stable and unstable.   The physical interpretation is quite clear.   The pendulum either oscillates about its lowest position ($x = 2n\pi$), or it has been given so high an initial velocity that it turns always in the same direction about the point of suspension.   In the latter case the angular displacement x increases without limit but the angular velocity fluctuates periodically about a certain average value.   It could be shown easily that the time required for the pendulum to reach its highest point (the unstable equilibrium positions for which $x = n\pi$, n odd) with velocity zero would be infinite, in accordance with our observations in the similar situation of the preceding example.

We observe that the equilibrium positions in our examples correspond to what we shall define later as singular points in the x,v-plane, and that the character of these singular points pretty well fixes the general character of the integral curves.   In the next chapter (in Sections 8, 9, and 10) we shall study such singularities in detail with the object of applying the knowledge so gained to the analysis of more difficult problems than those treated in the present chapter.   However, we also return in the next chapter to a few additional problems of the type treated here.

CHAPTER III

Free Oscillations with Damping and the Geometry of Integral Curves

1. The plan of this chapter

In this chapter we will be concerned with the differential equation

$$(1.1) \qquad \ddot{x} + \varphi(\dot{x}) + f(x) = 0$$

which contains a term provided by a damping force $-\varphi(\dot{x})$ in addition to the restoring force $-f(x)$. The equation (1.1) differs from the equation treated in the preceding chapter only through the occurrence of the damping term $\varphi(\dot{x})$, which makes explicit integrations not possible in general. However, we shall also re-examine some cases in which $\varphi(\dot{x}) \equiv 0$ by way of illustration of the more general approach of the present chapter as compared with the simpler approach possible in Chapter II.

The differential equation (1.1) arises, for example, in the case of a pendulum when damping forces are considered present. A differential equation of the same type occurs also, as we shall see, in problems concerning unsteady motions of synchronous electrical machinery and in a variety of other physical problems.

Since the time t does not occur explicitly in (1.1), it is possible to reduce the equation to one of first order by introducing $\dot{x} = v$, the velocity, as a new variable. One obtains in this way the equation

$$(1.2) \qquad \frac{dv}{dx} = \frac{-f(x) - \varphi(v)}{v} .$$

Because of the presence of the term $\varphi(v)$ in the right-hand side of (1.2) it is not possible to separate the variables, in general, to obtain the solution curves in the x,v-plane by explicit integration as was done in the preceding chapter. In spite of this, the geometric interpretation of (1.2) as an equation defining a field of directions in the x,v-plane can lead to useful information of a qualitative character even though the solution curves themselves cannot be obtained explicitly. By *solution curves* (or *integral curves*) *we mean*, of course, *curves which have everywhere the field direction defined by* (1.2).

There is some advantage in many cases in replacing the single equation (1.2) by the following equivalent pair of first order differential equations

(1.3)
$$\begin{cases} \dfrac{dv}{dt} = -f(x) - \varphi(v) \\[2mm] \dfrac{dx}{dt} = v \end{cases}$$

which yield a vector field with the components $(dx/dt, dv/dt)$; the field vector is always tangent to a solution curve and points along it in the direction of motion of the point $(x(t), v(t))$ in the x,v-plane with increasing t.

In Part A of the present chapter, comprising Sections 2 and 3, we give a number of examples indicating the value and usefulness of qualitative discussions of the integral curves of (1.2) or (1.3) in a number of physical problems. Among these problems are vibrations of a system with Coulomb damping, and of systems involving what are called self-excited oscillations.* In the course of this discussion the graphical method of Liénard is introduced, both for its own interest and with the object of showing that the qualitative discussion of the geometry of the integral curves of (1.2) and (1.3) can lead also to a quantitative discussion—for example, by graphical means.

We have so far considered the differential equation (1.2) as defining a field of directions in the x,v-plane. However, no such direction is defined at points where the numerator and denominator in the right-hand side of (1.2) vanish simultaneously. In terms of the equivalent equations (1.3) defining the vector field, we see that both components of the vector are zero at such points. *Such a point is called a singular point of the differential equation.* In our cases a singular point corresponds to a definite physical situation (when (1.1) is considered as an equation of motion of a mass), i.e., to a *position of equilibrium with velocity zero*, as one readily sees. Such singularities were encountered in the preceding chapter; in the case of the pendulum, for example.

---

* The latter problems are only touched upon here; in Chapter V we deal with them in great detail.

The character of the singularities in these cases was, in a way, a decisive factor in determining the qualitative nature of the solution curves, and hence it would seem important to study such singularities in detail. There is another important reason for making a study of singularities. As we know, closed curves in the x,v-plane correspond to periodic solutions $x(t)$ of the original second order differential equation and vice versa (cf., for example, Chapter II, Section 4). It is intuitively rather evident (and it has been proved rigorously under appropriate conditions on the functions involved) that such a closed integral curve always contains at least one singularity in its interior. In other words, the highly important case of closed solution curves in the x,v-plane is intimately connected with the occurrence of singularities.

Part B of this chapter is devoted to a detailed discussion of the possible types of singularities of the general first order differential equation $dv/dx = (P(x, v))/(Q(x, v))$ i.e. of the behavior of the solution curves of this equation in the neighborhood of a point where $P(x, v)$ and $Q(x, v)$ vanish simultaneously. It turns out, following the work of Poincaré, that a complete discussion of the types of singularities is possible and that criteria for distinguishing the various types can be given explicitly and rather simply in terms of P and Q. Finally, it is possible and useful to attach an integer number, called the *index*, to each of the types of singularities. A brief discussion of the notion of the index is also included in Part B of this chapter.

Finally, in Part C we make use of the ideas and results of Parts A and B to solve a variety of physical problems. These include a dynamical treatment of a problem in elastic stability and a treatment of the problem of the pull-out torque of a synchronous electric motor.

A. Geometrical and Graphical Discussion of Integral Curves

2. Geometrical discussion of the integral curves in a special case

The special case we have in mind is that of the free vibration of a mass with a nonlinear spring (more specifically, a hard spring) to provide a restoring force and operating in a medium which exerts a resistance proportional to the velocity, i.e. a medium providing viscous damping.   In other words, we consider the special case of (1.1) which arises when $\varphi(\dot{x}) = c\dot{x}$; we also assume $f(x)$ to be given by $f(x) = \alpha x + \beta x^3$ with $\alpha > 0$ and $\beta > 0$.   The equations (1.3) therefore become

(2.1)
$$\begin{cases} \dfrac{dv}{dt} = -(\alpha x + \beta x^3) - cv \\[2mm] \dfrac{dx}{dt} = v \end{cases}$$

in this special case.   In Figure 2.1 we indicate the closed solution curves of (2.1) for the case $c = 0$ (cf. Chapter II, Section 4), together with part of one integral curve for $c > 0$ as well; it is easy to deduce from (2.1) that the field vector at any point for $c > 0$ is in general

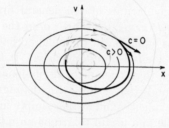

FIG. 2.1. Effect of positive damping on integral curves.

turned toward the interior of the closed solution curve for $c = 0$ through the point, as indicated in the figure.   The only exceptions are the points $v = 0$, $x \neq 0$ where the field vectors for $c > 0$ and $c = 0$ coincide; however, at these points it is not difficult to show that the solution curves for $c > 0$ cross the curve for $c = 0$ from the exterior to the interior with increasing t.   Since the solution curves for $c = 0$ are all closed curves, it seems quite clear (and in fact it could be proved with no difficulty) that all solution curves for $c > 0$ will tend to the origin as t increases.   In other words, *every motion would be damped out, and no periodic motion could exist.*   On the other hand, if c were negative (the case of so-called "negative damping") the displacement would increase indefinitely with t.

It is readily seen that the same conclusion holds as in the case $c > 0$ if a more general function $\varphi(v)$ of v is taken instead of cv provided only that $\varphi(v)$ has the same sign as v and is zero only for $v = 0$.

Again the motion defined by (2.1) cannot be periodic if $v\varphi(v)$ is always of one sign. For, suppose that (2.1) possesses a closed integral curve. We write (2.1) in the form (1.2), multiply by $v\,dx$, and integrate along the curve. The result is

$$\oint v\,dv + \oint \varphi(v)\,dx + \oint f(x)\,dx = 0.$$

This, in turn, yields

$$\oint \varphi(v)\,dx = 0, \quad \text{or} \quad \int_{t}^{t+T} v\varphi(v)\,dt = 0,$$

in which T is the period of the motion. But the last equation is impossible since $v\varphi(v)$ was supposed to be of one sign and does not vanish identically.

One sees, therefore, that quite a little insight can be gained about the motions governed by (1.1) through relatively simple geometrical discussions of the integral curves furnished by (1.2).

3. Liénard's graphical construction

In certain special cases of importance the first order differential equation (1.2) can be treated graphically in a very simple way by a method due to Liénard [25]. This method is most frequently used to deal with what are called self-excited oscillations (cf. Chapter V), but it is also applicable in other cases. The special cases in question are those in which $-f(x)$, the spring force, is linear in x; by introducing new variables which are appropriate multiples of the original variables it is easily seen that (1.2) can be written without loss of generality in the form

$$(3.1) \qquad \frac{dv}{dx} = \frac{-\varphi(v) - x}{v}.$$

Liénard's construction is indicated in Figure 3.1; its purpose is to obtain the field direction graphically at any point. The curve $x = -\varphi(v)$ is first plotted. To determine the field direction at any point $P(x, v)$, the procedure is as follows: From P a line is drawn parallel to the x-axis until it cuts the curve $x = -\varphi(v)$ at R. From R a perpendicular is dropped to the x-axis at S; the field direction at P is then orthogonal to the line SP. That the construction yields the correct field direction is seen at once from (3.1) and the fact that the slope of the line SP is $v/(x + \varphi(v))$.

One might use the Liénard construction to obtain an approximation to a solution curve through any given point in the following way: At the initial point the field direction is determined graphically in the manner indicated. The integral curve in the neighborhood of this point is replaced by a short segment of its tangent—that is, by a segment taken in the field direction. At the end point of this segment the field direction is determined once more by the graphical construction, and the process is repeated. In this way an approximation to the integral curve by a polygon is obtained which could be made

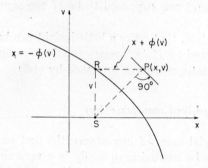

FIG. 3.1. Liénard's method for determining the field direction.

as accurate as desired by choosing the lengths of the sides of the polygon small enough.

There are a number of interesting cases in which the Liénard construction gives the integral curves immediately without the necessity of determining polygonal approximations. Consider, for example, the case of the linear free oscillations furnished by $\ddot{x} + x = 0$. In this case (3.1) is simply $dv/dx = -x/v$, $\varphi(v) \equiv 0$, the curve $x = -\varphi(v)$ is the v-axis, and hence the construction as indicated in Figure 3.1 shows the point S to be the origin for all points P. Hence the solution curves are, as we know they should be from the preceding chapter, circles with the origin as centers. We recall also that all of these circles are traversed by the point $(x(t), v(t))$ in the same time T, T being the period of the harmonic oscillation.

We turn next to the treatment of two problems which have been used by Meissner* [30] to illustrate the graphical method devised by him for the treatment of vibration problems. Consider first the case of a system with a mass and a linear spring subjected to solid friction, or Coulomb damping as it is also called. This is a friction force which is constant in magnitude but reverses its sign when the

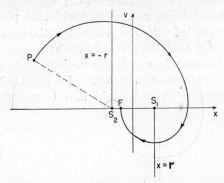

FIG. 3.2. Liénard's construction for the case of Coulomb damping.

velocity changes sign. The differential equation for the motion can be taken in the form

$$(3.2) \qquad\qquad \ddot{x} + \varphi(\dot{x}) + x = 0$$

with $\varphi(\dot{x})$ defined by

$$(3.3) \qquad\qquad \varphi(\dot{x}) = \begin{cases} r, & \dot{x} > 0 \\ -r, & \dot{x} < 0 \end{cases}, \qquad r > 0.$$

The curve $x = -\varphi(v)$ which figures in the Liénard construction is thus given by $x = -r, v > 0$ and $x = r, v < 0$ in the present case. Figure 3.2 indicates a typical solution curve. As we see at once, the integral curves consist of arcs of circles having S_1 and S_2 alternately as centers; on crossing the x-axis the center shifts from one to the other of these points. Since any point on a solution curve moves with increasing t in the direction of the arrows, it is clear that the

* For problems of the type under discussion here the writer believes that the method of Liénard is superior to Meissner's method; but it should be emphasized that Meissner's method, while more complicated, is also more general since it, unlike the method of Liénard, can be used to good advantage in cases in which the differential equation contains the time explicitly.

amplitude is decreased by the amount r between each successive pair of rest positions $v = 0$ until finally the mass comes to rest. The final rest position in the case shown in Figure 3.2 is denoted by F: this is, as in all cases, represented by the first point on the solution curve which falls on the segment S_1S_2, since the spring force in that case would have too small a value (cf. (3.2) and (3.3)) to overcome the friction force and hence no motion could take place. Suppose that the system is started with $x = A$, $v = 0$ at the initial time $t = 0$. One sees then readily that the number n of half-swings executed by the mass before it comes to rest is obtained by finding the nearest integer solution $x = n$ of the equation $rx = |A|$. The time required for the mass to come to rest could also be calculated without difficulty.*

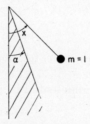

FIG. 3.3. Pendulum striking a partially inelastic wall.

Another example treated by Meissner using his method is that of a simple pendulum swinging against a wall inclined at angle α to the vertical. The impact of the mass of the pendulum on the wall is assumed to be inelastic in such a way that the value of the kinetic energy of the pendulum bob is decreased in a fixed ratio at each impact. The direction of the velocity is, of course, reversed at each impact. The linearized pendulum equation $\ddot{x} + x = 0$ is assumed. In the x,v-plane the solution curves are therefore circles with centers at the origin, as was pointed out above. In Figure 3.4 the solution

* The method of Meissner has been used by Ziegler [39] to deal with the problem of *forced oscillations* of a system with Coulomb damping. It might be of interest to note that infinite amplitudes can be built up by a periodic external force of proper frequency if the solid friction is not too great, as Ziegler shows. This is, of course, in strong contrast with the case of forced oscillations in linear systems with viscous damping, since the amplitude is always finite in the latter cases (cf. Chapter I).

curves for the problem of inelastic impact are indicated for $\alpha > 0$
(as in Figure 3.3), $\alpha = 0$, and $\alpha < 0$, when the kinetic energy of the
pendulum is assumed to be reduced by half its value at each impact.
The solution curves, which have discontinuities in the present case,
all tend when $\alpha \geq 0$ to the point $x = \alpha$, $v = 0$ as t increases as one
would expect; in other words, the pendulum bob tends to a position
of rest in contact with the wall. Meissner observed, however, that
there is an interesting qualitative difference between the case $\alpha > 0$
and the cases $\alpha \leq 0$. The fact is that the state corresponding to $x = \alpha$, $v = 0$ is obviously not approached at all for $\alpha < 0$, is approached
for $\alpha = 0$ only as the time tends to infinity, while it is "attained" for

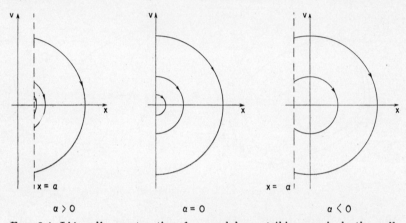

$$\alpha > 0 \qquad\qquad \alpha = 0 \qquad\qquad \alpha < 0$$

FIG. 3.4. Liénard's construction for pendulum striking an inelastic wall.

$\alpha > 0$ in finite time. That the time is infinite in the cases $\alpha \leq 0$ is
seen at once from the fact that the time required to traverse each of
the infinitely many circular arcs representing the whole motion is
equal to the central angle of that arc (in radians, of course) and the
fact that this angle is constant for the semicircular arcs in the case
$\alpha = 0$ and increases with each swing in the case $\alpha < 0$. In the case
$\alpha > 0$, the central angle can easily be shown to tend to zero for
successive swings in such a way that the sum of the angles converges
to a finite limit.

Finally, we indicate briefly the use of the Liénard construction in
studying what are called self-excited or self-sustained oscillations.
It is characteristic for these cases that the "damping force" $-\varphi(v)$

in (3.1) has the same sign as v for small values of v but the opposite sign for large values of v; in other words, this force acts in such a way as to tend to increase the amplitude of the oscillation when the velocity is small, but has the opposite effect when the velocity is large.   One might then expect that the interplay of these two effects of opposite tendency would lead to a motion which tends to a steady vibration. As we have already stated earlier, some of the important problems of this class will be treated in detail in a separate chapter (Chapter V) of this book.   We confine ourselves here to a reproduction in Figure 3.5 of the result of applying the Liénard construction to (3.1) for the case in which $-\varphi(v) = v - v^3/3$.   The curve $x = v - v^3/3$ is shown on the left, while a few integral curves obtained by step-wise applica-

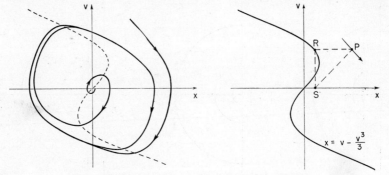

FIG. 3.5. A case of a self-sustained oscillation.

tion of the Liénard construction are shown on the right.   The latter curves indicate the general behavior of all integral curves: those which start near the origin spiral away from it while those which start far from the origin spiral toward it.   In addition, the actual carrying out of the construction convinces one that all integral curves spiral toward a single closed integral curve, thus indicating that *all motions of the system tend with increasing time to a single periodic motion.*

B. A Study of Singular Points

4. Singular points and criteria for their classification

The rest positions of equilibrium of the systems characterized by (1.1), (1.2), and (1.3) correspond to singularities of the first order differential equation (1.2). Such singularities have been of importance in our previous discussion but only in an incidental way. As we have already stated in Section 1 of this chapter, it is important to study such singularities in detail as an aid in the solution of various physical problems. Furthermore, we shall eventually have need of information about the singularities of more general first order differential equations than (1.2). We turn, therefore, to a discussion of the character of the solution curves in the neighborhood of an isolated singular point of the differential equation

$$(4.1) \qquad \frac{dv}{dx} = \frac{P(x, v)}{Q(x, v)}.$$

By a singular point (x_0, v_0) is meant, we repeat, a point for which $P(x_0, v_0) = Q(x_0, v_0) = 0$. (Note that a point for which $Q = 0$, $P \neq 0$ is not considered a singularity. In the vicinity of such a point we simply consider dx/dv instead of dv/dx. Note also that we may always shift the origin of the x,v-plane to the singularity, since dv/dx is invariant under such a change of coordinates.)

It has been shown by Poincaré [34] that the differential equation

$$(4.2) \qquad \frac{dv}{dx} = \frac{ax + bv + P_2(x, v)}{cx + dv + Q_2(x, v)},$$

in which the constants a, b, c, d are such that the determinant $\Delta = ad - bc \neq 0$ and in which P_2 and Q_2 vanish like $x^2 + v^2$ as $x,v \to 0$, has as its only singularities (at the origin $x = 0$, $v = 0$, of course) those of the much simpler equation

$$(4.3) \qquad \frac{dv}{dx} = \frac{ax + bv}{cx + dv}.$$

In addition, he showed that criteria for distinguishing the types of singularities of (4.2) can be derived solely in terms of the constants a, b, c, and d. If, however, $ad - bc = 0$ (which would occur, for example, if the developments of either P or Q were to begin with terms of order higher than the first) singularities of higher order and of quite different types from those obtainable from (4.3) can occur. However, the restriction $\Delta \neq 0$ is, in general, fulfilled in the cases of interest to us, so that we may confine our attention to (4.3). We shall assume without proof the result of Poincaré relating the singularities of (4.2) to those of (4.3), but we shall derive in detail the criteria for the classification of the singularities of (4.3). In the section to follow immediately we shall study the singularities of (4.3) for certain special values of the constants a, b, c, d by integrating (4.3) explicitly in these cases. In Section 6 we shall then prove that all cases of (4.3) (i.e. all cases with respect to changes of a, b, c, d) are reducible to these special cases by appropriate linear transformations, and at the same time we shall deduce the criteria for distinguishing the various types of singularities. The treatment used follows essentially the same lines as that in the book of Bieberbach [5].

5. Special cases of $dv/dx = (ax + bv)/(cx + dv)$

As indicated above, we study here a number of special cases of (4.3) in which the differential equation is easily solved explicitly. The singularity is, of course, at the origin of the x,v-plane.

1. $dv/dx = av/x$. The integral curves are easily found to be given by $v = v_0(x/x_0)^a$. If $a = 1$, the integral curves consist of all straight lines through the origin. If $0 < a < 1$ all integral curves pass through the origin and all are tangent to the v-axis there except the curve $v = 0$. If $a > 1$ all integral curves pass through the origin and all are tangent to the x-axis there except the curve $x = 0$. In all of these cases the origin is called a *nodal point*, or *node*. (See Figure 5.1a, 5.1b.) The situation is quite different if $a < 0$, $a = -k$, $k > 0$, say. The family of integral curves $vx^k = v_0x_0^k$ is asymptotic to the axes. Only the integral curves $x = 0$, $v = 0$ pass through the origin; all others pass by the origin. This type of singularity is called a *saddle point*. (See Figure 5.1d.) In the case of the pendulum, the unstable equilibrium positions correspond to singularities of this sort, as we have seen.

2. $dv/dx = -\mu^2x/v$. The integral curves are given by $v^2 + \mu^2x^2 = v_0^2 + \mu^2x_0^2$. All solution curves are ellipses with the origin as center. Only the degenerate solution curve $v^2 + \mu^2x^2 = 0$, i.e., $x = 0, v = 0$, passes through the origin. The singularity is called a *center* (cf. Figure 5.1e). The stable equilibrium points of the pendulum are of this type.

3. $dv/dx = (x + av)/(ax - v)$, $a \neq 0$. This equation is solved most easily by introducing polar coordinates $x = \rho \cos \theta, v = \rho \sin \theta$. In these variables the equation reduces to $d\rho/d\theta = a\rho$ and the integral curves are given by $\rho = ce^{a\theta}$; they are thus logarithmic spirals. The

singularity is called a *spiral point* (cf. Figure 5.1f). The x,v-curves for a free linear oscillation with damping are curves of this kind, and the origin of the x,v-plane is a spiral point of equilibrium, provided that the damping is less than the critical damping, i.e., $r^2 < k/m$. (See

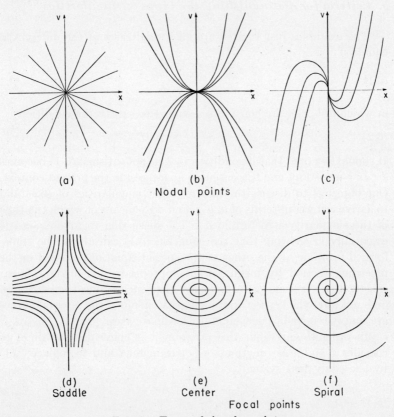

FIG. 5.1. Types of singular points.

Chapter I.) If the damping is greater than the critical damping, $r^2 > k/m$, the singularity at the origin of the x,v-plane is a node. If the damping is critical, $r^2 = k/m$, the equation for the x,v-curves can be transformed by a linear transformation into:

4. $dv/dx = (x + v)/x$. The integral curves of this equation are given by $v = xv_0/x_0 + x \log |x/x_0|$. All curves pass through the origin. The origin is again called a nodal point (Figure 5.1c).

In the following section we shall see that the differential equation (4.3) (and with it (4.2) if $\Delta \neq 0$) has no other types of singularities than the ones just obtained above.

6. Criteria for distinguishing the types of singularities

The examples just treated are all special cases of the differential equation

$$(6.1) \qquad \frac{dv}{dx} = \frac{ax + bv}{cx + dv},$$

in which a, b, c, d are constants assumed to satisfy the condition

$$(6.2) \qquad \Delta = ad - bc \neq 0.$$

It should be noted that if condition (6.2) is not satisfied (6.1) becomes $dv/dx = $ constant, and this case is of no interest in the present context. Our object is to discuss the nature of the singularities of (6.1) and to derive criteria in terms of a, b, c, and d by means of which the type of the singularity is determined. The discussion to follow consists essentially in showing that the equation (6.1) can always be transformed into one or the other of the special equations studied in the preceding section by means of an appropriate non-singular linear substitution on x and v. It is clear that such a linear substitution does not alter the type of the singularity: in effect the integral curves are simply referred to new oblique coordinate axes.

In order to carry out this program it is convenient at times to consider x and v as functions of a parameter t and to replace (6.1) by the equivalent system

$$(6.3) \qquad \begin{cases} \dot{v} = ax + bv, \\ \dot{x} = cx + dv. \end{cases}$$

The functions x and v are transformed into x_1, v_1 by the following linear transformation:

$$(6.4) \qquad \begin{cases} x_1 = \alpha x + \beta v, \\ v_1 = \gamma x + \delta v. \end{cases}$$

Our object is to seek out non-singular transformations (i.e. transformations such that $\alpha\delta - \beta\gamma \neq 0$) which lead to the following simple form for the differential equations for the new quantities v_1 and x_1 :

(6.5)
$$\begin{cases} \dot{v}_1 = \lambda_2 v_1, \\ \dot{x}_1 = \lambda_1 x_1, \end{cases}$$

with λ_1 and λ_2 certain constants; or, what comes to the same thing, $dv_1/dx_1 = (\lambda_2/\lambda_1)(v_1/x_1)$. This procedure is motivated to some extent by the fact that a majority of the special cases treated above in Section 5 were of this form. Since $\dot{x}_1 = \alpha \dot{x} + \beta \dot{v}$ and $\dot{v}_1 = \gamma \dot{x} + \delta \dot{v}$, we may write (6.5) in the form

(6.6)
$$\begin{cases} \alpha(cx + dv) + \beta(ax + bv) = \lambda_1(\alpha x + \beta v) \\ \gamma(cx + dv) + \delta(ax + bv) = \lambda_2(\gamma x + \delta v) \end{cases}$$

upon use of (6.3). Since the ratio of x and v is in general not constant, it follows that the equations

(6.7)
$$\begin{cases} \alpha(\lambda_1 - c) - \beta a = 0 \\ -\alpha d + \beta(\lambda_1 - b) = 0 \end{cases}$$

and the equations

(6.8)
$$\begin{cases} \gamma(\lambda_2 - c) - \delta a = 0 \\ -\gamma d + \delta(\lambda_2 - b) = 0 \end{cases}$$

must be satisfied for values of α, β, γ, δ not all zero in view of the assumed existence of the transformation (6.4). Consequently, λ_1 and λ_2 must be roots of the equation

(6.9)
$$\begin{vmatrix} c - \lambda & a \\ d & b - \lambda \end{vmatrix} = 0,$$

which is called the *characteristic equation*; it can also be written in the form

(6.10)
$$\lambda^2 - \lambda(b + c) - (ad - bc) = 0.$$

Since we assume always that $ad - bc \neq 0$ it follows that (6.10) has no vanishing root.

If the two roots λ_1 , λ_2 of the characteristic equation are unequal, it is readily seen that the determinant $\alpha\delta - \beta\gamma$ of the transformation (6.4) must be different from zero, as follows: Either ad or bc must be different from zero, since $ad - bc \neq 0$.   If, for example, $ad \neq 0$ we could obviously choose for β/α and δ/γ the values (cf. (6.7) and (6.8))

$$\beta/\alpha = -\frac{c - \lambda_1}{a}, \qquad \delta/\gamma = -\frac{c - \lambda_2}{a}$$

with $\beta/\alpha \neq \delta/\gamma$ since $\lambda_1 \neq \lambda_2$.   A similar statement can obviously be made in case $bc \neq 0$; it follows that a transformation (6.4) with $\delta\alpha - \beta\gamma \neq 0$ will exist so that v_1 and x_1 satisfy (6.5).

If the roots λ_1 and λ_2 are real as well as unequal, which will be the case if the discriminant $D = (b - c)^2 + 4ad$ of (6.10) is positive, we observe that a number of the previously discussed typical cases occur.   For example, if λ_2/λ_1 is negative, which in turn occurs if $\Delta = ad - bc$ is positive (cf. (6.10)), the singularity is a saddle point.   If, however, λ_2/λ_1 is positive, and hence Δ is negative, the singularity is seen to be a nodal point.   In the latter case also we observe from (6.5) that *a point on any solution curve moves into the singularity as t increases if the roots are both negative and away from it if the roots are positive*, i.e. *according to whether $b + c$ is negative or positive*.   In this way we can distinguish between stable and unstable nodal points.

If the roots λ_1, λ_2 are not real, and hence conjugate complex, we shall prove first that x_1 and v_1 are also conjugate complex, i.e. that they can be written in the form

$$\text{(6.11)} \qquad \begin{cases} v_1 = v_2 + ix_2 \\ x_1 = v_2 - ix_2 \end{cases}$$

with x_2 and v_2 both real.   Assuming for the moment that this is true we solve (6.11) for v_2 and x_2 to obtain

$$\text{(6.12)} \qquad \begin{cases} v_2 = \dfrac{1}{2}(x_1 + v_1) = \alpha_1 x + \beta_1 v \\[2mm] x_2 = \dfrac{i}{2}(x_1 - v_1) = \gamma_1 x + \delta_1 v \end{cases}$$

upon using (6.4); since v_2 and x_2 are assumed real, it follows that α_1, β_1, γ_1, δ_1 are real and in addition the determinant of these coefficients does not vanish since the transformation (6.12) is the result of applying successive transformations each of which has a non-vanishing determinant.   For dv_2/dx_2 we have in any case the following differential equation, obtained from (6.12) and (6.5):

$$\frac{dv_2}{dx_2} = \frac{(\dot{x}_1 + \dot{v}_1)}{i(\dot{x}_1 - \dot{v}_1)} = \frac{(\lambda_1 x_1 + \lambda_2 v_1)}{i(\lambda_1 x_1 - \lambda_2 v_1)}$$

$$(6.13) \qquad = \frac{i(\lambda_2 - \lambda_1)x_2 + (\lambda_1 + \lambda_2)v_2}{(\lambda_1 + \lambda_2)x_2 - i(\lambda_2 - \lambda_1)v_2}$$

$$= \frac{Ax_2 + Bv_2}{Bx_2 - Av_2},$$

in which A, B are obviously real constants since λ_1 and λ_2 are complex conjugates. As a consequence, real functions x_2 and v_2 satisfying our relations exist and the validity of the transformation (6.12) is established. Furthermore, the constant A in (6.13) cannot be zero, since the roots λ_1 and λ_2 are unequal and conjugate complex, and hence the differential equation (6.13) again belongs to one of the typical cases discussed in the preceding section, i.e. to case 3 in which the integral curves are spirals, unless $B = \lambda_1 + \lambda_2$ is zero, in which case the integral curves are circles with center at the origin. Upon transforming back to the x,v-plane by using (6.12) one sees that the integral curves are spirals if $\lambda_1 + \lambda_2 \neq 0$ and ellipses if $\lambda_1 + \lambda_2 = 0$. Furthermore, it can also be shown that a point on a spiral moves into the origin with increase of t if $\lambda_1 + \lambda_2$ is negative and away from the origin if $\lambda_1 + \lambda_2$ is positive, i.e. according to whether the real part of the characteristic roots is negative or positive.

Finally, we consider the case in which the roots of the characteristic equation are equal, i.e. $\lambda_1 = \lambda_2 = \lambda$, and hence real. It is possible in this case that the equations (6.7) and (6.8) are satisfied identically; but this requires that $a = d = 0$ and that $b = c$ so that $dv/dx = v/x$ and the integral curves consist of the straight lines through the origin. If (6.7) and (6.8) are not satisfied identically, we can make use of the root λ in order to determine non-vanishing values for α and β and with this choice transform the original differential equation into the following form:

$$(6.14) \qquad \frac{dv_1}{dx_1} = \frac{a_1 x_1 + b_1 v_1}{x_1},$$

the values of γ and δ in (6.4) being any values linearly independent of

α and β.   For the differential equation (6.14) the characteristic equation is

$$\begin{vmatrix} 1 - \lambda & a_1 \\ 0 & b_1 - \lambda \end{vmatrix} = 0,$$

but since it is assumed to have equal roots it follows that $b_1 = 1$. If a_1 were zero, (6.14) would have the form $dv_1/dx_1 = v_1/x_1$ and the integral curves would again be the straight lines through the origin. If $a_1 \neq 0$ we may introduce a new transformation $v_2 = \dfrac{1}{a} v_1$, $x_2 = x_1$, the determinant of which does not vanish, and obtain in place of (6.14) the equation

$$\frac{dv_2}{dx_2} = \frac{x_2 + v_2}{v_2},$$

which is the case treated in 4 of the preceding section so that the singularity is once more a nodal point.   Thus if $\lambda_1 = \lambda_2$ the singularity is a nodal point.   If λ is negative, a point on any integral curve moves into the origin as t increases, away from it if λ is positive.

The results of the above discussion can be summed up in the following table, in which we refer to a singularity as stable or unstable according to whether a point on any integral curve moves into the singularity or not with increasing t:

I. $(b - c)^2 + 4ad > 0$ $\begin{cases} \text{(A) Node} & \text{if } ad - bc < 0 \\ \text{(B) Saddle} & \text{if } ad - bc > 0 \end{cases}$ $\begin{cases} \text{Stable} & \text{if } b + c < 0 \\ \text{Unstable} & \text{if } b + c > 0 \end{cases}$

II. $(b - c)^2 + 4ad < 0$ $\begin{cases} \text{(A) Center if } b + c = 0 \\ \text{(B) Spiral if } b + c \neq 0 \end{cases}$ $\begin{cases} \text{Stable} & \text{if } b + c < 0 \\ \text{Unstable} & \text{if } b + c > 0 \end{cases}$

III. $(b - c)^2 + 4ad = 0$ Node.   (Case 4) $\begin{cases} \text{Stable} & \text{if } b + c < 0 \\ \text{Unstable} & \text{if } b + c > 0 \end{cases}$

This completes our discussion of the classification of singularities of first order differential equations, except for one additional remark. We have already stated that the singularities obtained in the present section are the same as those for the general first order equation:

$$\frac{dv}{dx} = \frac{ax + bv + P_2(x, v)}{cx + dv + Q_2(x, v)}$$

in case $ad - bc \neq 0$ and P_2 and Q_2 vanish to at least second order at the origin. Poincaré also showed that the *criteria* for the classification of the types of singularities also remain unchanged in this more general case, with one exception: the condition $b + c = 0$ no longer suffices to distinguish between a center and a spiral in the case $D < 0$; the higher order terms P_2 and Q_2 must be taken into account for this purpose.

7. The index of a singularity

For some purposes it is useful to assign a number, called the index, to the various types of singularities of first order differential equations treated above. The definition of the index of a singularity, as given by Poincaré [34], is as follows: Consider a simple closed curve $C(t)$ (i.e., a closed curve without double points) which passes through no singularities and which has at most one singularity in its interior. The angle $\theta(t)$ which the field vector $(dx/dt, dv/dt)$, defined by the differential equations

$$\begin{cases} \dfrac{dv}{dt} = P(x, v) \\[2mm] \dfrac{dx}{dt} = Q(x, v) \end{cases}$$

at the points of $C(t)$, makes with the positive x-axis is taken in such a way that $\theta(t)$ is continuous. On making one complete circuit around C in the counterclockwise direction the angle $\theta(t)$ changes by an amount $2j\pi$, j a positive or negative integer (or zero), since the field vector returns to its original position after completing the circuit. *The number j is called the index of the singularity.* If $j = -1$, for example, this means that the field vector makes one complete revolution in the clockwise sense when C is traversed once counterclockwise; if $j = 0$ the field vector may oscillate but it does not make a complete rotation.

It is clear that this definition of the index is not appropriate unless the index has the same value for every curve C enclosing the given singularity (and no others). This is, however, the case. One proves it along the following lines: Suppose C' is another simple curve enclosing the singularity. Upon deforming C into C' in a continuous manner it is very plausible (and can be proved rigorously) that the angle θ and with it the index also vary continuously, but on the other hand since the index is an integer it must remain constant.

It is easy to see geometrically, on the basis of this definition, what the index is for each of the singularities discussed in Section 3

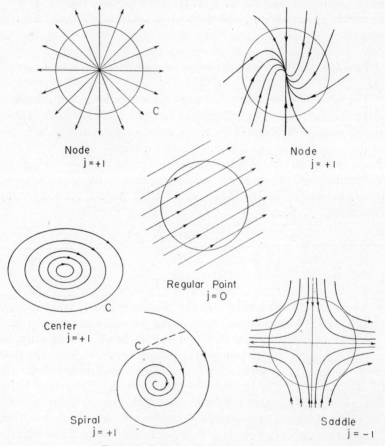

Node
$j = +1$

Node
$j = +1$

Regular Point
$j = 0$

Center
$j = +1$

Spiral
$j = +1$

Saddle
$j = -1$

FIG. 7.1. Indices of singular points.

above (cf. Figure 7.1): nodes, centers, and spiral points all have the index $+1$, while the saddle point has the index -1. In the case of a center, the curve C may be chosen as one of the closed integral curves. At a regular point (i.e. a point that is not a singular point) one sees that the index is zero. It is possible to define other higher types of singularities for which j is different from ± 1, but since these cases are not of importance for our later discussions we shall not consider them here.

The notion of *index of a closed curve* C is defined in the obvious way by means of the change in the angle $\theta(t)$ upon making a circuit around C, and it is of interest to know that the index of a closed curve containing a finite number of singularities is equal to the algebraic sum of their indices. The reason for this can be made intuitively clear without difficulty as follows: The interior of the closed curve C is divided into a number of regions each of which contains one of the singularities in its interior (cf. Figure 7.2). We recall that the index of a singularity was defined as the change in the angle $\theta(t)$ of the field vector in making a circuit in the counterclockwise direction around

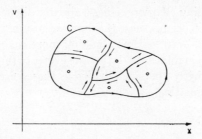

Fig. 7.2. Index of a simple closed curve.

a curve enclosing the singularity. The sum of all the indices inside C could therefore be obtained by adding the angle changes for all of the curves. But the angle changes over the segments of the boundary curves inside C occur always twice with opposite signs, since these curve segments are traversed twice in opposite directions (again cf. Figure 7.2). Hence the contributions to the total change in angle furnished by the interior curve segments cancel out and we see therefore that the index of C is the sum of the indices of the singularities in its interior.

A knowledge of the nature and distribution of the singularities is not enough to determine the qualitative character of the solution curves of a first order differential equation, but such information can be highly useful in certain important special cases. For example, suppose that there is a closed solution curve C without double points or other singularities. Since the tangent vector to such a curve turns through the angle 2π in the positive sense on making one circuit of the curve, it follows that *the sum of the indices of all singularities inside any closed solution curve free of singularities is* $+1$. It follows, therefore, that there must be at least one singularity inside such a closed solution curve. Furthermore, if the singularities are all of the types discussed here, in which the indices 'are $+1$ or -1, we see that the number of saddle points must be one less than the number of other types of singularities. This observation will be very important for the discussion of certain problems in Chapter V.

C. Applications Using the Notion of Singularities

8. Free oscillations without damping

In the preceding chapter we have already discussed in some detail the motions which occur in case the differential equation characterizing them is $\ddot{x} + f(x) = 0$, without making more than incidental use of the notion of a singularity. It is, however, useful and illuminating to consider explicitly the singularities in these cases also. We take the differential equation

$$(8.1) \qquad \frac{dv}{dx} = \frac{-f(x)}{v}$$

with $f(0) = 0$ so that the origin is a singularity, and write

$$(8.2) \qquad f(x) = a_1 x + \frac{a_2}{2} x^2 + \cdots, \qquad a_1 \neq 0.$$

We assume $a_1 \neq 0$, i.e., that the spring stiffness does not vanish for $x = 0$; otherwise the determinant Δ (cf. (6.1) and (6.2)) would vanish and the singularity would be of a higher order than we wish to consider. Upon introduction of the potential energy $F(x)$ through

$$(8.3) \qquad F(x) = \int f(x)\,dx = \frac{a_1}{2!} x^2 + \frac{a_2}{3!} x^3 \cdots$$

the energy integral becomes

$$(8.4) \qquad v^2/2 + F(x) = h$$

or

$$(8.5) \qquad v^2/2 + \left[\frac{a_1}{2!} x^2 + \frac{a_2}{3!} x^3 \cdots \right] = h$$

in which h is the total energy of the system. As we noted in the preceding chapter, the singularities correspond to rest positions of equilibrium, since the conditions $f(x) = 0$, $v = 0$ characterize these points. In terms of the potential energy $F(x)$ the singularities therefore occur for values of x satisfying $F'(x) = 0$. Since $dv/dx = (-a_1x + \cdots)/v$, we have, in the notation of Section 6, $a = -a_1$, $b = 0$, $c = 0$, $d = 1$, and the table at the end of Section 6 shows the singularity to be a saddle if $a_1 < 0$ and a center if $a_1 > 0$. From (8.3) we see therefore that saddle points correspond to maxima of the potential energy $F(x)$ and minima to centers. If the assumption

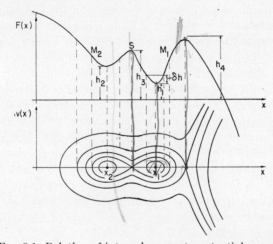

FIG. 8.1. Relation of integral curves to potential energy.

$a_1 \neq 0$ were to be given up, the singularity at $x = 0$ would be of higher order, as we have already observed. However, the nature of the singularity would be the same if $F(0)$ were a relative minimum or maximum, as can readily be seen from (8.5). But if $F'(0) = 0$ while $F(0)$ is neither minimum nor maximum the singularity is in general of a type different from any we have discussed hitherto.

As an example of the kind of conclusions that can be drawn with regard to the character of the motions for a given $F(x)$, consider Figure 8.1 in which a special function $F(x)$ is plotted together with the x,v-curves for various values of the energy constant h. We observe first of all that the change of origin from $x = 0$ to $x = x_1$ obtained by setting $\xi = x - x_1$ replaces (8.1) by $dv/d\xi = [-f(\xi +$

$x_1)]/v$ and the discussion above now applies to a neighborhood of $\xi = 0$. We consider first the case $h = h_1$ corresponding to the lowest minimum of $F(x)$ in the range we consider. It is clear that the only motion possible is the state of rest with $x = x_1$, $v = 0$, since $v^2 < 0$ for all other x. If h is changed slightly to $h = h_1 + \delta h$, $\delta h > 0$, the energy curve is a closed curve with $(x_1, 0)$ in its interior, corresponding to the fact that the singularity for $x = x_1$ is a center. As h is gradually increased new periodic motions can arise as soon as h passes the new minimum point at $x = x_2$, $h = h_2$ on $F(x)$. So far all motions might reasonably be considered to be stable since slight changes in initial conditions cause only slight changes in the resulting motions. However, when $h = h_3$ is reached a quite different type of curve appears, corresponding to the fact that $F(x)$ has h_3 as a maximum at point S. A solution curve with a saddle point arises, as indicated in the figure. In this particular case the solution curve with the saddle point is a closed curve with abscissas delimited by the points M_1 and M_2, as one can readily see. When h is increased farther to h_4 another saddle singularity arises. The corresponding solution curve has a closed loop to the left, but is open on the right. For values of h between h_3 and h_4 there exist closed solution curves which enclose a saddle point and two centers in their interiors. We have here a confirmation of a general result dealt with in the preceding section that a closed solution curve free of singularities must contain in its interior one more center than saddle so that the sum of the indices of the singularities inside it is $+1$.

The above discussion also makes it rather clear that the solution curves through the saddle singularities would in general be decisive in fixing the qualitative character of all the solution curves, since they separate the plane into regions in which the solution curves behave in different ways.

9. Wire carrying a current and restrained by springs

The above discussion can be extended to the somewhat more general motions determined by the differential equation

$$(9.1) \qquad\qquad \ddot{x} + f(x, \lambda) = 0,$$

i.e. to motions in which the spring force depends upon a parameter λ in addition to the displacement x. The solution curves and the character of their singularities depend then upon λ; in particular, it is possible for one or more of the singularities to change in type suddenly when λ passes through certain values, and consequently that the character of the possible motions of the system may change radically when λ passes the transition values. We shall consider in this and the next section two special physical problems which illustrate some of the possibilities in this connection.

The first problem* is that of the motion of a current-carrying conductor restrained by springs and subjected to a force from the magnetic field due to another infinitely long fixed parallel wire, as

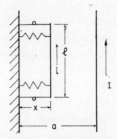

FIG. 9.1. Elastically restrained current-carrying wire in a magnetic field.

indicated in Figure 9.1. The differential equation governing the motion of the wire is

$$(9.2) \qquad m\ddot{x} + k\left(x - \frac{\lambda}{a - x}\right) = 0.$$

The parameter λ is given by $\lambda = 2Iil/k$, and k is the spring constant. The term $k\lambda/(a - x)$ is the force of attraction set up by the magnetic fields due to the current in the wires. In the present case we have

$$(9.3) \qquad \frac{dv}{dx} = \frac{k}{m}\frac{x^2 - ax + \lambda}{v(a - x)}.$$

The singularities are the points $(x_1, 0)$, $(x_2, 0)$ with x_1 and x_2 the roots of $x^2 - ax + \lambda = 0$, i.e.

$$(9.4) \qquad x_{1,2} = \frac{a}{2} \pm \sqrt{\frac{a^2}{4} - \lambda}.$$

Upon setting $\xi_i = x - x_i$ we can replace (9.3) by the equation

$$(9.5) \qquad \frac{dv}{d\xi_i} = \frac{k}{m}\frac{\xi_i(\xi_i + x_i - x_j)}{v(a - x_i - \xi_i)} = \frac{k}{m}\frac{(x_i - x_j)\xi_i + \cdots}{v(a - x_i) + \cdots}$$

* This problem is treated in the book of Minorsky [31].

in which x_i , x_j refer to the roots furnished by (9.4) and the dots refer to quadratic terms which are to be neglected near $\xi_i = 0$, $v = 0$ in determining the character of the singularities at $\xi_1 = 0$ and $\xi_2 = 0$, i.e. at $x = x_1$ and $x = x_2$.

For $\lambda \neq 0$, which we assume, the numerator on the right-hand side of (9.3) does not vanish for $x = a$, and (9.3) can be written in the form $v dv = dx/g(x)$ with $g(x)$ a function vanishing to the first power in $x - a$ at $x = a$. Hence v tends to infinity as x approaches a. We shall therefore restrict x to values less than a. There are then three cases to be discussed, depending on the nature of the roots $x_{1,2}$ as determined by the value of λ: (a) $\lambda < a^2/4$; both roots are real and positive with $x_1 > x_2$. (b) $\lambda = a^2/4$; both roots have the value $a/2$. (c) $\lambda > a^2/4$; the roots are not real and hence no equilibrium positions exist. The force due to the magnetic field is, in fact, always larger than the spring force in this case.

In the first case there are two singularities at $x = x_1$ and $x = x_2$ with $x_1 > x_2$, both of which lie between $x = 0$ and $x = a$. From (9.5) we see from the criteria developed in Section 6 that the singularity is a saddle at $x = x_1$ since $(x_1 - x_2)$ and $(a - x_1)$, the coefficients of ξ_1 and v, are both positive, and is a center at $x = x_2$ since $(x_2 - x_1)$ is negative and $(a - x_2)$ is positive.

The same conclusion regarding the character of the singularities can also be obtained readily in the present case, just as in the preceding case, by considering the potential energy $F(x)$, which is in this case given by

$$(9.6) \qquad F(x) = kx^2/2 + k\lambda \log (a - x),$$

as one finds immediately from $f(x) = k[x - \lambda/(a - x)]$. In Figure 9.2 we indicate $F(x)$ together with the solution curves of (9.3) for the cases $\lambda < a^2/4$ and $\lambda = a^2/4$. In the former case the solution curves are closed curves surrounding the point $(x_2 , 0)$ until the energy constant is so large that the solution curve contains the saddle singularity at $(x_1 , 0)$. Since $F(x)$ always increases as x decreases below $x = x_2$, we see that the solution curve through the saddle point forms a loop on the left. For still larger values of the total energy there are two distinct types of open solution curves corresponding to two different types of non-periodic motions. In other words, the motion of the wire is periodic only if the initial velocity and displacement of the wire are not too large. When $\lambda = a^2/4$ the two singularities coalesce to form a singularity of higher order (cf. (9.5) in which the coefficient $(x_i - x_i)$ of ξ_i is now zero and hence the numerator on the right-hand side is quadratic in ξ_i). The curve for $F(x)$ has a point of inflexion with a horizontal tangent at $x = a/2$. The solution curve through $(a/2, 0)$ has a cusp at this point; there are no closed solution curves and consequently no periodic motions. In case (c) when $\lambda > a^2/4$, there are no singularities and hence no periodic motions,

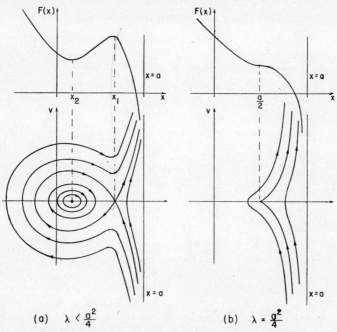

(a) $\lambda < \dfrac{a^2}{4}$ (b) $\lambda = \dfrac{a^2}{4}$

FIG. 9.2. Integral curves for wire in a magnetic field.

and the solution curves have the same general appearance as the curves to the left of the singularity indicated in Figure 9.2(b).

The interpretation of these results for the physical problem is easily given: If the spring is sufficiently weak, no vibrations occur and the mass simply moves toward the fixed wire without oscillating. On the other hand, if the spring is stiff the motion will be oscillatory if the initial position of the mass is not too near to the fixed wire nor its initial velocity too high; otherwise, the mass again will tend eventually to move always toward the fixed wire. This behavior is of course conditioned by the fact that the force of attraction due to the magnetic field becomes infinite when the fixed wire is approached.

10. Elastic stability treated dynamically

If a slender straight elastic rod is subjected at its ends to compressive forces along the axis of the rod it is well known that the straight equilibrium position will not be stable unless the compressive forces are kept under a certain critical value, beyond which the column

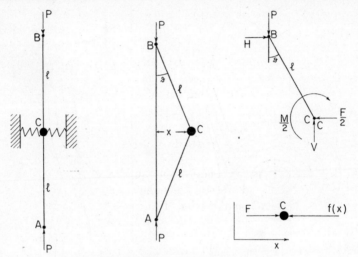

FIG. 10.1. Analogue of an elastic column.

bends or buckles. This problem is usually treated in the theory of elasticity from a purely static point of view by inquiring simply for the magnitude of the loads at the ends of the column at which bent states of equilibrium may exist in addition to the straight state, and defining as the critical buckling load the smallest load for which such a bent state of equilibrium can occur.

In this section we shall consider a very much simplified version of the stability problem for the column, but treat it dynamically instead of statically. The simplification results through considering an elastic system having only one degree of freedom instead of the column, which as an elastic continuum has infinitely many degrees of freedom. In Figure 10.1 we indicate the system to be treated.

It consists of two light rods pivoted together at point C and free to slide along a vertical line at their other ends A and B. At point C the pivot is assumed to carry a particle of mass m. Compressive forces P, as indicated, act at the ends A and B of the "column" along the vertical line through A and B. At point C springs are provided which act to produce a sidewise restoring force $f(x)$ which depends upon the displacement x. In addition, the two rods are assumed not to rotate relative to each other at C without constraint, but rather to react on each other through a properly attached coil spring, for example (not shown in the figure), which exerts a restoring moment M proportional to the angle through which the rods turn relative to each other.* The forces acting on one of the rods (the forces on the other rod are the same by symmetry) and on the mass m are indicated in the figure. Since we neglect the mass of the rod the following conditions on the forces acting are obtained from statics:

(10.1) $$P = V,$$

(10.2) $$-\frac{M}{2} + Vl \sin\theta - \frac{Fl}{2} \cos\theta = 0.$$

The first equation is the condition of equilibrium of the forces in the vertical direction, and the second is the equation of moments about point B. The equation of motion of the mass m is

(10.3) $$m\ddot{x} = F - f(x).$$

For the restoring moment M provided by the coil spring at C we have $M = 2k_1\theta$ with k_1 a constant, while for $f(x)$, the lateral spring force, we assume the behavior to be given by $f(x) = \alpha x + \beta x^3$, with α and β both positive constants. Equation (10.3) then takes the form

(10.4) $$m\ddot{x} - 2\frac{Px - k_1 \text{ arc sin } x/l}{\sqrt{l^2 - x^2}} + (\alpha x + \beta x^3) = 0$$

upon using (10.1) and (10.2) to eliminate V and $x = l \sin\theta$ to eliminate θ. We now assume that the sidewise displacement x is so small compared with l that powers of x/l above the third may be ignored compared with lower powers, and thus replace (10.4) by the following equation

(10.5) $$m\ddot{x} + (\alpha + 2k_1/l^2 - 2P/l)x + (\beta + 4k_1/3l^4 - P/l^3)x^3 = 0.$$

* In this way we simulate the *bending* stiffness of a continuous elastic column.

We observe that one solution of our problem is given by $x \equiv 0$, corresponding to the straight equilibrium position. The question is whether this solution is stable or not when the pressure P is increased.

It is perhaps of interest to interpolate at this point the treatment of the stability problem on a purely static basis, as one does in the theory of elasticity. By dropping the acceleration term we obtain from (10.5) the following equilibrium condition

$$(10.6) \qquad a_1 x + a_3 x^3 = 0,$$

with a_1 and a_3 given by

$$(10.7) \qquad a_1 = \alpha + 2k_1/l^2 - 2P/l,$$
$$(10.8) \qquad a_3 = \beta + 4k_1/3l^4 - P/l^3.$$

In the theory of elasticity one also usually restricts oneself to displacements x so small that nonlinear terms in x can be ignored; in this case the equilibrium condition takes the form $a_1 x = 0$, with a_1 defined by (10.7). Consequently $x = 0$ unless $a_1 = 0$, and the latter equation in turn furnishes from (10.7) the following "critical" value for P:

$$(10.9) \qquad P_{\text{crit.}} = \alpha l/2 + k_1/l$$

at which the column would buckle, since the equilibrium becomes indifferent for this value of P in the sense that x is now arbitrary.

We turn now to the dynamical treatment of the problem, which yields the same value for $P_{\text{crit.}}$ but determines it through an argument which is perhaps more convincing. In accordance with our usual practice we replace $\ddot{x}$ by $v(dv/dx)$ and obtain the first order equation

$$(10.10) \qquad \frac{dv}{dx} = \frac{1}{m} \frac{-a_1 x - a_3 x^3}{v},$$

with a_1 and a_3 as defined in (10.7) and (10.8). In the x,v-plane the origin is an equilibrium point; and it is the character of the singularity there which decides whether the state $x = 0$ is stable or not. As we have seen in the previous examples, the sign of a_1 is decisive for this: if a_1 is positive, the singularity is a center, while it is a saddle point if a_1 is negative. The two cases are indicated in Figure 10.2. If $a_1 > 0$, a slight disturbance from the equilibrium position results in a small oscillation, but if $a_1 < 0$, the motion departs widely from the equilibrium position upon the slightest disturbance. From (10.7) and (10.9) we conclude therefore that the column is stable if $P < P_{crit.}$ and unstable if $P > P_{crit.}$, so that the static determination of $P_{crit.}$ yields the correct transition value from a stable to an unstable equilibrium position.

In Figure 10.2 the curves for the case $P > P_{crit.}$ were drawn assuming that a_3 is positive (which would be the case if β were large enough, for example), so that two equilibrium positions would occur for $x = \pm \sqrt{a_3/(-a_1)}$. One finds readily that the corresponding singularities are centers.

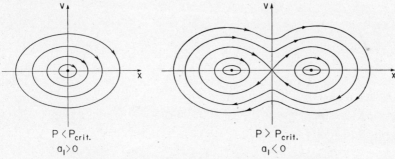

$P < P_{crit.}$ $P > P_{crit.}$

$a_1 > 0$ $a_1 < 0$

Fig. 10.2. Stable and unstable cases of the column.

It is interesting to consider the effect of viscous damping on our mechanical system. In equation (10.4) we would have an additional term $c\dot{x}$, $c > 0$, on the left side of the equation, and equation (10.10) would be replaced by

$$(10.11) \qquad \frac{dv}{dx} = \frac{1}{m} \frac{-a_1 x - cv - a_3 x^3}{v}.$$

This equation cannot be solved by using the energy integral, but through consideration of the nature of the singularities of the equation it is not difficult to discuss the character of the solution curves. In the first place we observe that the equilibrium positions are not altered by including the damping term, since $v = 0$ in these positions. In the stable case when $a_1 > 0$ one finds readily, with reference to the table at the end of Section 6, that the singularity is a stable nodal point if the damping constant c is sufficiently large and is a stable spiral point otherwise. For this purpose we must identify $-a_1$ with a,

$-c$ with b, take $c = 0$, and set $d = 1$ in using the table of Section 6. In either case, the displacement x tends with increasing t to the rest position $x = 0$. However, if $a_1 < 0$ or, in other words, if $P > P_{\text{crit.}}$, the singularity is readily found to be a saddle point, just as in the case without damping. As we would expect (and could also easily show from the criteria for singularities), the two other singularities, which

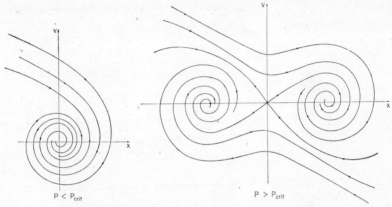

FIG. 10.3. Effect of damping on motion of the column.

were centers in the case of no damping, now become stable spiral or nodal points. The solution curves in the x,v-plane now appear as in Figure 10.3. In the unstable case, we observe that all motions tend to one or the other of the two stable equilibrium positions. Once more the solution curves which contain the saddle point separate the plane into regions in one of which all solution curves tend to one of the two stable singularities, and in the other region to the second stable singularity.

The dynamic treatment of our stability problem for a simplified model of an elastic column thus leads to a completely satisfactory explanation of the buckling phenomena. In the case of the continuous elastic column a dynamic treatment of the stability problem would also be more satisfying than the present static treatment, but the difficulties in carrying out such a treatment would not be small. Certainly the above treatment would be out of the question, since it is in principle confined to systems with one degree of freedom.

11. The pendulum with damping proportional to the square of the angular velocity

In the remainder of this chapter we shall deal with cases in which the occurrence of damping forces plays an essential role. As a first example we take the case of a pendulum immersed in a medium which exerts a force proportional to the square of its velocity and in a direction opposite to the velocity. The differential equation for the pendulum in this case can therefore be written in the form

$$(11.1) \qquad \ddot{x} + c\dot{x} \,|\, \dot{x} \,| + k \sin x = 0.$$

The quantity x is the angle of swing measured from the lower equilibrium position. Instead of (11.1) we consider, as usual, the first order equation

$$(11.2) \qquad \frac{dv}{dx} = \frac{-k \sin x - cv \,|\, v \,|}{v}.$$

The singularities, corresponding to the equilibrium positions, are at $x = n\pi$, n a positive or negative integer. At $x = 0$, the right-hand side of (11.2) takes the form $(-kx + \cdots)/v$ and the singularity is therefore either a center or a spiral: we have here the exceptional case mentioned at the end of Section 6 in which the criteria for the singularity fail to distinguish between these two cases. However, we know that the singularity at $x = 0$ for $c = 0$ (that is, the case of no damping) is a center, and we have seen in Section 2 of the present chapter that the singularity then becomes a stable spiral point when damping is present. At $x = \pi$ the singularity is a saddle, as we know from earlier discussions, since this singularity is not affected by an additional quadratic term $cv \,|\, v \,|$. It is therefore clear that the singularities at $x = n\pi$, $v = 0$ are stable spiral points if n is even and saddle points if n is odd. The solution curves of (11.2) are readily seen to appear as shown in Figure 11.1. Every motion tends to a stable equilibrium position.

The curves of Figure 11.1 can be obtained explicitly in the present case by integrating (11.1) in the following way: We introduce $v = \dot{x}$ and $y = v/\sqrt{k}$ in (11.1) to obtain

$$(11.3) \qquad y\frac{dy}{dx} + cy \,|\, y \,| + \sin x = 0,$$

which can in turn be written in the form

$$(11.4) \qquad \frac{d(y^2)}{dx} + 2cy^2 = -2 \sin x \quad \text{for} \quad y > 0,$$

and

$$(11.5) \qquad \frac{d(y^2)}{dx} - 2cy^2 = -2 \sin x \quad \text{for} \quad y < 0$$

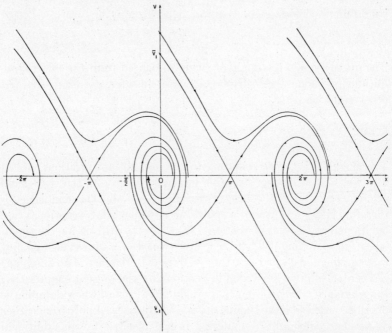

Fig. 11.1. Integral curves for pendulum with quadratic damping.

Equations (11.4) and (11.5) are first order linear differential equations with constant coefficients for the function y^2; they are readily solved to yield

$$(11.6) \qquad y^2 = c_1 e^{-2cx} + \frac{2}{1 + 4c^2} \cos x - \frac{4c}{1 + 4c^2} \sin x, \qquad y > 0,$$

$$(11.7) \qquad y^2 = c_2 e^{2cx} + \frac{2}{1 + 4c^2} \cos x + \frac{4c}{1 + 4c^2} \sin x, \qquad y < 0.$$

The quantities c_1 and c_2 are arbitrary integration constants.

We now consider a special problem which foreshadows a type of consideration that will be important in the sections to follow immediately. Suppose that the pendulum is given an impulse when in the position $x = 0$ so that it acquires the initial angular velocity $\bar{v}$. If $\bar{v}$ is sufficiently small the pendulum will oscillate about $x = 0$ without making a full revolution about its support, but if $\bar{v}$ is large enough the pendulum will go "over the top" one or more times before finally oscillating about the stable equilibrium position. In terms of the phase curves in Figure 11.1 this means that a curve starting at $(0, \bar{v})$ will finally spiral down on one of the singularities $x = n\pi$ (n an even integer), which one it will be depending on the value of $\bar{v}$.

Our problem is to give the ranges of values of the initial velocity $\bar{v}$ within which the corresponding x,v-curve will spiral down on a given singularity, or, *in other words*, to *give the range of initial velocities within which a motion with a specified number of full revolutions will occur.* From Figure 11.1 it is quite obvious what must be done to solve the problem: *The upper branches of the solution curves through the saddle points $x = \pi, 3\pi, 5\pi, \cdots, v = 0$ must be continued backwards* (i.e. for decreasing values of t) until they cross the v-axis at points $\bar{v}_1, \bar{v}_3, \cdots$. If then $\bar{v}$ lies in the range $\bar{v}_n < \bar{v} < \bar{v}_{n+2}$ the pendulum will make $(n + 1)/2$ complete revolutions before coming to rest.

The values $\bar{v}_n$ delimiting the various ranges of $\bar{v}$ can be determined explicitly from (11.6) with no difficulty in the present case: We need only fix c_1 so that $y = 0$ for $x = n\pi$ (n odd) and then calculate $\bar{y} = \bar{v}/\sqrt{k}$ for $x = 0$. This yields for c_1 the value $c_1 = 2e^{2cn\pi}/(1 + 4c^2)$ and for $\bar{v}_n^2$ the value

$$(11.8) \qquad \bar{v}_n^2 = \frac{2}{1 + 4c^2}(1 + e^{2cn\pi}).$$

As we have seen, the pendulum will execute $(n + 1)/2$ full revolutions if $\bar{v}$ lies in the range $\bar{v}_n < \bar{v} < \bar{v}_{n+2}$. The impulse required to cause the pendulum to execute n revolutions therefore goes up exponentially with n.

12. The pendulum with viscous damping

In this section we treat essentially the same problem as in the previous section, except that the damping is assumed to be viscous damping, i.e. it is assumed to be proportional to the first power of the angular velocity rather than to its square. We have therefore to study the differential equation

$$(12.1) \qquad \ddot{x} + c\dot{x} + k \sin x = 0, \qquad c > 0.$$

Again we wish to determine the smallest initial impulse that must be given the pendulum when in the position $x = 0$ in order that it should make a full revolution before tending again to the rest position $x = 0$ as the time increases. Evidently, one could solve the problem by investigating the solutions $x(t)$ of (12.1) (graphically or otherwise) which satisfy the initial conditions $x(0) = 0$, $\dot{x}(0) = v_0$ for various values of v_0 until the value of v_0 is found for which x just reaches the angle π. It can, however, be treated in a simpler way to which one is led naturally when one interprets the problem in the x,v-plane and works backward from a singularity, as in the problem treated in the preceding section. This problem has practical importance when interpreted in terms of the unsteady operation of the synchronous alternating current motor, as we shall see in the next section.

In the previous section we were able to solve the corresponding problem easily by explicit integration of the differential equation, but that is not possible with (12.1). We shall therefore indicate three different methods for the numerical solution of our problem which are capable of furnishing any desired degree of accuracy. As a basis for these solutions we take, as usual, the first order differential equation resulting from (12.1) by introduction of $v = \dot{x}$ as new variable:

$$(12.2) \qquad \frac{dv}{dx} = \frac{-cv - k \sin x}{v}.$$

By setting $y = v/\sqrt{k}$, $\lambda = c/\sqrt{k}$ we obtain in place of (12.2) the following equation containing only one parameter:

$$(12.3) \qquad \frac{dy}{dx} = \frac{-\lambda y - \sin x}{y}.$$

Our problem is to determine the value of y for $x = 0$ so that x will just attain the value π. The solution curves for (12.3) are shown in Figure 12.1: qualitatively they do not differ from those shown in Figure 11.1 for the case of quadratic damping. The singularity at $x = 0$ is a spiral point (or a nodal point if λ is large enough), and at $x = \pi$ it is a saddle point, as we know from our previous discussions, or can verify by the criteria given in Section 6. Our object will

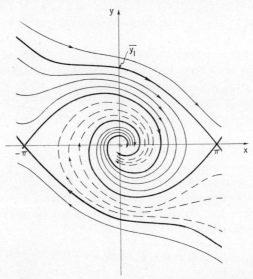

FIG. 12.1. Integral curves for pendulum with viscous damping.

clearly be achieved *once the point marked $\bar{y}_1$ in Figure 12.1 is determined,* i.e. *the point of intersection with the y-axis of the solution curve drawn backward from the saddle at $x = \pi$.* It is the determination of this solution curve which we propose to carry out numerically in three different ways in order to illustrate the kinds of procedures which may be used in similar cases.

The first method we use is the power series method. If a new independent variable $\xi = x - \pi$ is introduced in place of x in (12.3), this equation becomes

$$(12.4) \qquad y\left(\frac{dy}{d\xi} + \lambda\right) = -\sin(\xi + \pi) = \sin \xi.$$

A power series solution valid near $\xi = 0$ is desired; hence we write with undetermined coefficients:

$$(12.5) \qquad y = a_1\xi + a_3\xi^3 + a_5\xi^5 + \cdots.$$

(It can be seen from the differential equation that y is an odd function of ξ.) If this series is inserted in (12.4) the result is

$$(a_1\xi + a_3\xi^3 + \cdots)(\lambda + a_1 + 3a_3\xi^2 + 5a_5\xi^4 + \cdots) = \xi - \frac{\xi^3}{3!} + \cdots,$$

from which, by equating coefficients of like powers of ξ, we can obtain a set of equations for the coefficients a_i of the series. The first such equation is the following quadratic in a_1 :

$$(12.6) \qquad a_1(a_1 + \lambda) = 1,$$

with the real roots

$$(12.7) \qquad a_1 = (-\lambda \pm \sqrt{\lambda^2 + 4})/2.$$

Once one of the two roots has been chosen for a_1, the other coefficients are determined successively by linear equations; for instance, a_3 satisfies the equation

$$(12.8) \qquad a_3(\lambda + a_1) + 3a_3a_1 = -1/6.$$

This furnishes a verification of our statement that the point $x = \pi$, $y = 0$ is a saddle singularity since exactly two solutions enter it. The two solutions enter the singularity with slopes of different sign; but clearly our problem requires that we follow backward to $\xi = -\pi$ the one with the negative slope (cf. Figure 12.1).

We proceed to carry the solution through in the special case for which $\lambda = 0.1$. For this value of λ the first terms in the power series are

$$(12.9) \qquad y = -1.0512\xi + .0406\xi^3 - .0005\xi^5 + \cdots.$$

Even for $\xi = 1$ the error in using only the first term in the series would probably be less than 5 percent. In this particular numerical example we could make use of the power series to determine $y(-\pi) = \bar{y}_1$.

In later examples of a similar type it will not be possible to use the power series over the full range of values of ξ. Hence we indicate methods of continuing the solution from the singularity in the present case by a different method which can be used in other cases.

The solution could be pursued beyond the range in which (12.9) is accurate by any of a number of approximation methods. We choose first the method of finite differences. The interval $-\pi \le \xi \le 0$ is broken into a number of equal intervals of length h as shown in

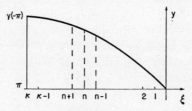

FIG. 12.2. Approximation to integral curves by finite difference method.

Figure 12.2. At each net point n we approximate the derivative by the formula

$$(12.10) \qquad \left. \frac{dy}{d\xi} \right|_n \cong \frac{y_{n-1} - y_{n+1}}{2h}$$

in which the right-hand side is the average of the forward and backward difference quotients. Instead of seeking a function $y(\xi)$ which satisfies (12.4) at all points of $-\pi \le \xi < 0$ we require that the difference equations obtained from (12.4) by replacing $dy/d\xi$ by (12.10) be satisfied at every net point. The difference equation for our problem is thus

$$\frac{y_{n-1} - y_{n+1}}{2h} = -\lambda + \frac{\sin \xi_n}{y_n}$$

or

$$(12.11) \qquad y_{n+1} = 2h \left(\lambda - \frac{\sin \xi_n}{y_n} \right) + y_{n-1}.$$

With this form of the difference equation we must know two values for y_i initially in order to proceed with the solution. Clearly we must take $y_0 = y(0) = 0$, while $y_1 = y(-h)$ *can be computed from the power series* (12.5) *if h is not too large.* From (12.11) we then find y_2.

Once y_2 is known (12.11) furnishes the value of y_3, etc. The accuracy obtainable by this method of course depends on the smallness of h; it could be shown that the values of y_i converge to the exact values as $h \to 0$.

We carry the calculations through for $h = \pi/4$. The value of y_1 as computed from (12.9) is $y_1 = y(-\pi/4) = .806$. From (12.11) the value of $y_2 = y(-\pi/2)$ is obtained; it is $y_2 = 1.53$. In the same way we obtain $y_3 = 1.99$ and finally $y_4 = \bar{y}_1 = y(-\pi) = 2.24$. Using three terms of the power series (12.9) would furnish the value $y(-\pi) = 2.20$.

Still another method of approximation can be applied in the present case. Since λ is small we can approximate the solution by iteration, assuming as a first approximation $y^{(0)}$ the solution of (12.3) for $\lambda = 0$ which vanishes at $x = \pi$, that is by

$$(12.12) \qquad y^{(0)} = \sqrt{2}\,\sqrt{1 + \cos x}.$$

The next approximation $y^{(1)}$ is obtained from (12.3) by inserting $y^{(0)}$ in the right-hand side to yield the following equation for $y^{(1)}$

$$\frac{dy^{(1)}}{dx} = -\lambda - \frac{\sin x}{y^{(0)}}$$

with $y^{(1)}(\pi) = 0$. The result is

$$(12.13) \qquad y^{(1)} = -\lambda(x - \pi) + \sqrt{2}\,\sqrt{1 + \cos x}.$$

In the case treated above $\lambda = 0.1$. Hence (12.13) yields for $y(0)$ the value $y(0) = 0.1(\pi) + 2 = 2.31$. The results of the three different methods of numerical solution differ by only a few percent.

13. Description of the operation of alternating current motors

The methods developed in this chapter have as one of their most striking applications the solution of the problem of pull-out torques of synchronous motors. It was therefore thought worth while to give a brief description of the mode of operation of these motors before taking up the pull-out problem in the next section.

Rotating alternating current motors all have one feature in common. A magnetic field which rotates in space is present. One of the simplest ways to obtain such a field is indicated in Figure 13.1. Two identical coils of rectangular shape are placed with their planes at right angles. The two coils are supplied with current I from a two-phase alternating source such that

$$I_{(1)} = I_m \cos \omega t \text{ and } I_{(2)} = I_m \sin \omega t,$$

that is the currents are 90° out of phase. At the time $t = 0$ the field around the wires would then be as indicated in Figure 13.1(c). At $t = \pi/2\omega$ (or, as we shall also say, 90° later) the field would encircle the coil (2) and not the coil (1).   If the fields furnished by each of the coils happened to be parallel fields of constant intensity everywhere, it would follow easily that the resulting field due to the superposition of both fields would also be a parallel field of the same constant intensity turning at angular velocity ω. With only two coils such a

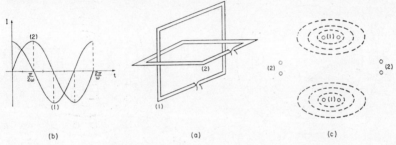

FIG. 13.1. Method of producing a rotating magnetic field.

field of constant intensity could hardly be achieved, but by taking n coils at angles $2\pi/n$ apart and using a polyphase current source a fairly close approximation to a rotating field of constant form and intensity can be obtained. The main point for our considerations here is that there is a magnetic field which rotates in space; that it may fluctuate somewhat in rotating is not of any great consequence for our purposes.

So far we have considered only what is called the armature of the motor, that is, a winding fixed in space and so arranged that it produces a magnetic field which rotates in space. The second essential element in a motor is an element free to rotate inside the armature and generally called the rotor. Different types of motors differ mainly in the design of the rotor. We consider here only two possibilities: the induction or squirrel cage motor, and the synchronous motor.

The rotor of a squirrel cage induction motor consists of a series of copper bars attached to the surface of a cylinder (see Figure 13.2); the bars are all connected together electrically at the ends. Such a motor functions as follows. Suppose that the rotor were at rest in the presence of the rotating magnetic field provided by the armature. The bars of the rotor would be cut by the magnetic lines of force and hence currents would be induced in them; as a consequence forces would be exerted at right angles to the bars and a torque would result. Such a torque would, furthermore, cause the rotor to turn in the sense of the rotating field. If no load were applied to the motor (and if losses in the motor itself were neglected) the steady state would be one in which the rotor turns at the same speed as the

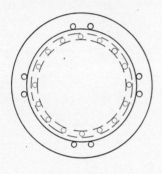
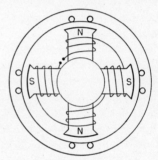

Induction Motor   Synchronous Motor

FIG. 13.2. Rotors of two types of motors.

rotating field, the so-called synchronous speed, and no currents would flow in the rotor bars. If the motor were overcoming a constant resisting torque, the rotor would turn at a speed less than synchronous speed; the "slip" of the rotor relative to the field would lead to induced currents and consequently to a torque just sufficient to overcome the load on the motor plus the resistances of various kinds in the motor itself. If $\dot{x}$ represents the angular velocity of slip of the rotor relative to the rotating field, the torque on the rotor is given approximately by $c\dot{x}$, with c a constant.

The principle of operation of the synchronous motor is quite different. In this type of motor the rotor consists of a series of wound electromagnets with iron cores, as indicated in Figure 13.2. The number of poles is just double that of the separate windings in the armature. The windings on the poles are so arranged that they are alternately of opposite polarity when the windings are connected to a source of direct current. For such a motor a separate source of *direct current* for the rotor and a source of alternating current for the armature are needed. The synchronous motor is not self-starting. If the rotor were at rest when its field and that of the armature were excited it is clear that the torque on the rotor would alternate rapidly in sign so that the time average of the torque over even quite short intervals of time would be zero. However, if the rotor were, by some means or other, made to rotate at or near the speed of the rotating field, the two fields would "lock" and a torque would be exerted. The principle of operation is that of a magnet mounted on a shaft which is caused to rotate by turning another magnet about it. When the motor is carrying no load, the two fields are locked so that they are parallel; when the motor is overcoming a torque the rotor field is turned at an angle x to the armature field so that a torque proportional to sin x is developed (of course, under steady conditions, the rotor continues to turn at synchronous speed). It is as though the two fields were coupled together elastically. Of course, the torque which can be carried by the rotor is limited by the strength of the fields; if the torque is made too large the rotor will "skip a pole," in the electrical engineer's terminology, and the motor will "fall out of step" and cease to operate.

The synchronous motor can be made self-starting by adding copper bars to the pole faces so that the motor can also function to greater or less degree as an induction motor in the manner described above. In such a case the operation of the motor is governed (to a first approximation at least) by the same differential equation as for the pendulum with viscous damping:

$$(13.1) \qquad \ddot{x} + c\dot{x} + k \sin x = L,$$

in which the term $\ddot{x}$ represents the inertia torque due to the mass of the rotor and its connected load, the term $c\dot{x}$ arises from the torque due to the "slip" of the squirrel cage winding relative to the rotating field, the term $k \sin x$ is due to the angle between the fields of the rotor and armature, and L represents the torque of the external load on the motor. The angle x is measured from an axis which rotates with the synchronous speed of the rotating electrical field.* Actually,

---

* The angle x is now not the space angle but is rather the angle measured in so-called electrical measure: 2π electrical radians correspond to the space angle between two alternate poles of the same polarity on the rotor of the synchronous motor.

the differential equation (13.1) is somewhat oversimplified since the factor c varies somewhat in practice and the term $k \sin x$ should be supplemented in general by terms involving sines and cosines of $2x$, but the methods of attacking (13.1) presented here would serve also in these more complicated cases.

The problem concerning the pendulum with damping solved in the preceding section can be interpreted in terms of the operation of a synchronous motor, as follows: Suppose that the motor is operating under no load, i.e. $L = 0$, and with the rotor "locked" with the rotating field so that $x = 0$. At a certain instant an impulsive torque is applied to the rotor so that the angular velocity $v = \dot{x}$ suddenly changes from the value zero to the value $\bar{v}$. The problem is to find the smallest value $\bar{v}_1$ such that x will just reach the value π, i.e. such that the magnetic poles of the rotor pass from a position in which they are in line with poles of *opposite polarity* in the rotating magnetic field to a position in which poles of the *same polarity* are in line. The latter position is on physical grounds obviously an unstable equilibrium position. Any increase in $\bar{v}$ above $\bar{v}_1$ would then cause the rotor to "skip a pole" and the motor would in most cases "fall out of step" and cease to operate. The methods of the preceding section would yield the value of this "pull-out impulse" for any given motor once the values of c and k in (13.1) appropriate to the motor are given.

14. Pull-out torques of synchronous motors

A more general and in practice a more important "pull-out" problem for the synchronous motor than the problem discussed in the preceding section is the following: Up to the time $t = 0$ the motor is operating at constant speed under a steady torque L_0, i.e. x is a constant x_0 which satisfies the relation

$$(14.1) \qquad k \sin x_0 = L_0 .$$

At $t = 0$ a sudden additional torque L_1 is applied but is then held constant so that the total torque has the constant value $L_0 + L_1$. The problem, called the problem of the pull-out torque, is to *determine the maximum additional torque L_1 which can be applied for each value L_0 of the original steady torque without causing the rotor to reach the* *nearest unstable equilibrium position, with probable resulting operational failure.* We observe that $L_* = L_0 + L_1$ must always be less than k in order that an equilibrium state $x_* = $ constant with torque L_* be possible at all, since the relation $k \sin x_* = L_*$ must be satisfied.

It is of interest to interpret this problem also in terms of the pendulum with viscous damping, even though the problem is then a rather artificial one from the physical point of view. We assume that external moments L are applied to the pendulum. In Figure 14.1 we indicate the stable equilibrium position $x = x_0$ (marked A) corresponding to $L = L_0$. Position B is the stable equilibrium position corresponding to the total torque $L_* = L_0 + L_1$, while C (vertically above B) is clearly the unstable equilibrium position which corre-

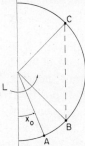

FIG. 14.1. Pendulum with external applied moment.

sponds to the torque L_*. Since the torque L_* is larger than the equilibrium value L_0 (we assume here that L_0 and L_1 are both positive) for $x = x_0$, where the pendulum is initially at rest, a motion of the pendulum from A toward C will ensue when the torque L_1 is suddenly added to L_0. If L_1 is not too large the pendulum will tend eventually to the stable equilibrium position B corresponding to the torque L_* without attaining the unstable equilibrium position C. If, however, L_1 exceeds a certain value $\overline{L}_1 = \overline{L}_* - L_0$ (which depends upon L_0 and the constants in (13.1)) the pendulum will pass the point C and will (as we shall see later) continue on to make a complete revolution. The determination of $\overline{L}_1$ (or, what is the same thing, $\overline{L}_*$) for each value of L_0 constitutes the analogue of the pull-out problem for the synchronous motor.

Before proceeding to discuss the solution of the problem, it is convenient to introduce in (13.1) a new independent variable τ in place of t through the relation

(14.2) $$t = \tau/\sqrt{k}$$

and parameters λ and γ defined by

(14.3) $$\lambda = c/\sqrt{k}, \quad \gamma = L/k.$$

The new differential equation is

(14.4) $$\frac{d^2 x}{d\tau^2} + \lambda \frac{dx}{d\tau} + \sin x = \gamma$$

which contains only two parameters. We note that the condition

(14.5) $$|\gamma| < 1$$

must be satisfied, since otherwise no state of equilibrium would be possible.

In terms of these new quantities the pull-out problem is formulated mathematically as follows: Up to the time $t = 0$ the motion furnished by (14.4) is the state of stable equilibrium $x = x_0 = $ constant, with $\gamma = \gamma_0 = $ constant; hence x_0 and γ_0 satisfy the condition $\sin x_0 = \gamma_0$ with x_0 the smallest positive root of this equation. At the instant $t = 0$ the value of γ is suddenly changed from the value γ_0 to a new constant value γ_* ; we set $\gamma_* = \gamma_0 + \gamma_1$, so that γ_1 is the change in γ. The numerical value of γ_* is taken always to be less than unity: $|\gamma_*| < 1$, in order that an equilibrium state x_* satisfying $\sin x_* = \gamma_*$ can exist. The pull-out problem then requires the determination of the largest value $\bar{\gamma}_1$ (or, of $\bar{\gamma}_* = \gamma_0 + \bar{\gamma}_1$) for each γ_0, such that the solution $x(t)$ of (14.4) with $\gamma = \gamma_*$, which at $t = 0$ satisfies $x = x_0$, $\dot{x} = 0$ with $\sin x_0 = \gamma_0$, will not pass the value π—or, in terms of the pendulum indicated in Figure 14.1, that the pendulum will just not "go over the top" and make a full swing. Actually, it is not necessary to require that the value π be attained: it is sufficient to require that x should just reach the nearest unstable equilibrium position, which occurs for $x < \pi$ in general, since x would go on to attain the value π if γ_1 were increased the slightest amount beyond $\bar{\gamma}_1$, as we shall see.

As we know, it is not necessary to work with the differential equation (14.4); instead, one can work to better advantage with the first order differential equation

(14.6) $$\frac{dv}{dx} = \frac{-\lambda v - \sin x + \gamma}{v}$$

which results from (14.4) upon introducing $v = dx/d\tau$ as new independent variable.* The method of solution, which follows the lines of our discussions in the two preceding sections, is to take for γ the value $\bar{\gamma}_*$ (satisfying (14.5), of course), and begin at the unstable saddle singularity $x = x_s$, $v = 0$ corresponding to position C in Figure 14.1 and hence satisfying the equation $\sin x_s = \gamma_*$. *The solution curve is then followed backward to position A, that is, to the first point $x = \dot{x}_0$ where v becomes zero.* From $\sin x_0 = \gamma_0$ the value of γ_0 is determined and with it the value of $\bar{\gamma}_1$ from $\bar{\gamma}_* = \gamma_0 + \bar{\gamma}_1$. From $\gamma = L/k$ (cf. (14.3)) the corresponding values of L_0, $\bar{L}_1$, and $\bar{L}_*$ are at once determined. *The essence of this approach to the pull-out problem lies in the idea of starting with the critical value $\bar{\gamma}_*$ of γ_* and working backward to the corresponding initial value γ_0*, instead of the other way around, which would be the more obvious approach. However, it should be observed that the recommended procedure of assuming a value for $\bar{\gamma}_*$ and then calculating a critical value for γ_0 does not directly solve the original pull-out problem, which requires that γ_0 be prescribed and the corresponding critical value $\bar{\gamma}_*$ be calculated. But we shall show later that *the same pairs of values of γ_0 and $\bar{\gamma}_*$ are coordinated by fixing values of $\bar{\gamma}_*$ first and determining the corresponding critical values for γ_0, as by working the other way around.*

The method of attacking the problem directly by starting with an assumed value for γ_0 has been used by Lyon and Edgerton [28] to solve the pull-out problem for various values of the constant λ characterizing the amount of damping (cf. (14.3)) and for initial values γ_0 of γ covering the range $0 \leq \gamma_0 < 1$. Their method was to solve the second order differential equation (14.4) graphically by use of the integraph at the Massachusetts Institute of Technology. In this method, a whole family of curves $x(\tau)$ must be drawn to determine the critical value of the torque for each given λ and γ_0. The procedure is to start with the initial values $x(0) = x_0$, $\dot{x}(0) = 0$ and follow the solution curves $x(\tau)$ of (14.4) for a fixed value of λ and various values of γ until $x(\tau)$ either exceeds the value π or until it becomes apparent that $x(\tau)$ approaches a value less than π. The value of γ at the transition is then the value we call $\bar{\gamma}_*$. The procedure outlined above, in which equation (14.6) is used instead of (14.4), is obviously much more efficient (only one solution curve of a

---

* Minorsky [31], Ch. VII, discusses the work of Vlasov on the operation of the synchronous motor along essentially the same lines. However, the pull-out problem is not formulated.

first order equation need be found for each γ_0 and λ) and, as we shall see, it is also not difficult to carry out numerically in such a way as to yield a complete solution of the problem. Of course, it should at the same time be made clear that our method of attacking the present problem cannot be applied except to special problems, while the integraph can be used quite generally to solve initial value problems.* Also, Lyon and Edgerton obtain x directly as a function of τ, while our method would require an additional numerical integration to achieve this.

We turn, then, to the solution of the pull-out problem following the ideas sketched out above. The simplest case to consider is that in which the damping factor λ is zero. In this case (14.6) becomes

$$(14.7) \qquad \frac{dv}{dx} = \frac{-\sin x + \bar{\gamma}_*}{v}$$

which can be integrated explicitly. The singularities are located on the x-axis where $x = \arcsin \bar{\gamma}_*$; it is readily checked that they are alternately centers and saddles, as one expects on physical grounds, the lowest positive root x_c of $x = \arcsin \bar{\gamma}_*$ being a center and the next largest $x_s = \pi - x_c$ a saddle. We assume here, as always in this section, that $\bar{\gamma}_* < 1$. The x,v-curves are shown in Figure 14.2. We are interested in the upper branch of the curve which enters the saddle singularity $x = x_s$, $v = 0$ and, in particular, in the point $x = x_0$ where this curve crosses the x-axis again. We need only integrate (14.7) to obtain $v(x)$ under the condition $v = 0$, $x = x_s$; the result is

$$(14.8) \qquad \frac{v^2}{2} = \cos x - \cos x_s + \bar{\gamma}_*(x - x_s),$$

in which $x_s = \pi - \arcsin \bar{\gamma}_*$ and the smallest positive value for arc sin $\bar{\gamma}_*$ is taken. The point x_0 is then located by calculating the value of x from (14.8) for $v = 0$. The points x_c, x_s, and x_0 are shown in Figure 14.2, which gives the results of accurate calculations for the case $\bar{\gamma}_* = .8$. As noted above, this procedure yields directly the solution of a problem which is different from the original pull-out problem. However, we shall not show the equivalence of the two problems here since we intend to prove it a little later for the general case in which λ is not necessarily zero.

* For example, the method of Lyon and Edgerton could be used to solve the pull-out problem in cases when the external torque is not constant in the time, while our method would not be applicable.

If $\lambda \neq 0$, i.e., if the damping constant is not zero, it is not possible to integrate (14.6) explicitly. Also, the singularities change in that centers become spirals (or nodes); saddles, however, remain saddles, as one can verify from the criteria at the end of Section 6. Typical x, v-curves are shown in Figure 14.3. The pull-out problem is solved here in the same way as above by taking a fixed value for $\bar{\gamma}_*$ and determining the corresponding value of x_0. We proceed to show briefly

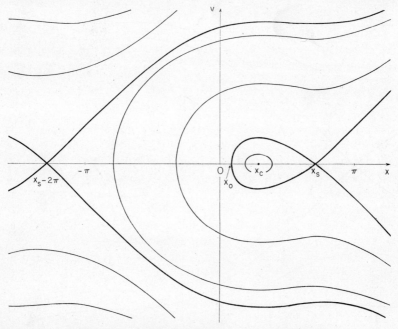

FIG. 14.2. Pull-out problem for zero damping.

that the critical value of γ which would result if x_0 (or γ_0) were held fixed and γ were varied would be the value $\bar{\gamma}_*$ assumed in our procedure. On account of the general form of the integral curves, as indicated in Figure 14.3, *it is sufficient for this purpose to show that the point x_0 moves to the right if $\bar{\gamma}_*$ is increased*, since this would clearly mean that the critical value of x_0 for a given value of $\bar{\gamma}_*$ (as calculated by our procedure) is a value for which an increase in $\bar{\gamma}_*$ would lead to a displacement beyond the unstable equilibrium position. Hence $\bar{\gamma}_*$ is the critical value belonging with the value x_0. We first observe that the upper branch of the integral curve from x_0 to x_s (cf. Figure 14.3) is given by $v = v(x)$ with v a single-valued positive function of x. Consider then two different arcs $v = v_1(x)$ and $v = v_2(x)$ corresponding to two different values $\bar{\gamma}_{*1}$ and $\bar{\gamma}_{*2}$ of $\bar{\gamma}_*$ with $\bar{\gamma}_{*2} > \bar{\gamma}_{*1}$, which extend from the points x_{01}, x_{02} to x_{s1}, x_{s2} respectively. In what follows we are concerned only with the arcs between these pairs of points. We must show that the point x_{01} belonging to $\bar{\gamma}_{*1}$ lies to the left of the point x_{02}. This is shown in the following way:

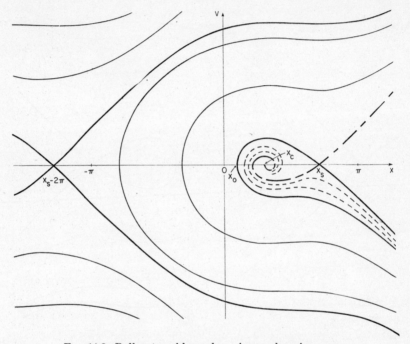

FIG. 14.3. Pull-out problem when viscous damping occurs.

We observe first that $x_{s2} < x_{s1}$ since these values satisfy $x_{si} = \pi - \text{arc}$ $\sin \bar{\gamma}_{*i}$ and $\bar{\gamma}_{*2} > \bar{\gamma}_{*1}$ was assumed. The integral curve $v = v_2(x)$ for $\bar{\gamma}_{*2}$, which begins at x_{s2}, therefore starts *under* the arc $v = v_1(x)$ extending from x_{01} to x_{s1} (cf. Figure 14.4, in which the arc $v = v_1(x)$ is shown dotted). We show next that the arc $v_2(x)$ cannot intersect the arc $v_1(x)$ as follows: If it did, there would be a *first* such intersection, say at point P, and it is clear that the slope of the arc $v_2(x)$ at this point could not be algebraically greater than the slope of the arc $v_1(x)$ at the same point. On the other hand, since $v_1(x)$ and $v_2(x)$

both satisfy (14.6), and v and x are the same for both solutions at point P, it is readily seen that $dv_2/dx > dv_1/dx$ since v is positive and γ is greater for v_2 than for v_1. Hence no intersection P can exist, and the arc $v_2(x)$ cannot cut the arc $v_1(x)$ except possibly at x_{01}. The latter possibility can also be ruled out easily by considering x as a function of v in the neighborhood of the point x_{01} and comparing values of dx/dv on the two arcs through use of (14.6); one would find that $v_2(x)$ would necessarily lie above $v_1(x)$ near x_{01}, in contradiction with the fact proved above that $v_2(x)$ lies under $v_1(x)$. We see therefore that the arc $v_2(x)$ lies entirely under the arc $v_1(x)$ and must cross the x-axis to the right of x_{01}, which proves our statement.

Calculations to determine the value of x_0 for given values of λ and $\bar{\gamma}_*$ can be carried through by the methods used in Section 12

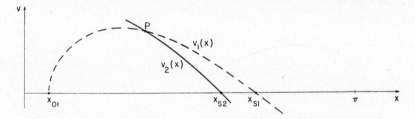

Fig. 14.4. Solution curves for different values $\bar{\gamma}*$ of.

above for the pendulum with viscous damping. We introduce in place of x a new variable $\xi = x - (\pi - x_c)$ where $\sin x_c = \bar{\gamma}_*$, $x_c < \pi/2$. The value $\xi = 0$ thus corresponds to the saddle singularity $x_s = \pi - x_c$. We wish to determine the value ξ_0 of ξ (and hence of $x = x_0$) for which v becomes zero once more. (See Figure 14.3.) As in Section 12 a solution $v(\xi)$ of (14.6) in the form of a power series valid near $\xi = 0$ can be obtained:

$$(14.9) \qquad v = a_1\xi + a_2\xi^2 + a_3\xi^3 + \cdots.$$

Once a numerical value for $\bar{\gamma}_*$ has been taken, it is not difficult to compute the successive coefficients; the calculation is of exactly the same type as was carried out in Section 12. However, the series (14.9) cannot be expected to furnish the quantity we seek, the value of ξ for which $v(\xi) = 0$, since $dv/d\xi$ is infinite there and (14.9) would therefore diverge for such a value. It is, then, imperative in this case to use some other method in conjunction with the power series. One could use the method of finite differences in the same way as in Section 12, except that it would not be very accurate near ξ_0 where $dv/d\xi$ becomes infinite. However, this difficulty can be overcome by interchanging the roles of v and ξ near this point, i.e., by considering ξ as a function of v and proceeding with finite differences in the usual way. Of course the solution curve for v as a function of ξ should be carried somewhat beyond the maximum $(dv/d\xi = 0)$ before taking v as independent variable. In this manner the solutions of the pull-out problem were calculated for the following three cases:

a) $\lambda = .022$, $\bar{\gamma}_* = 0.8$ with the results: $\gamma_0 = .22$, $x_0 = .22$, and $\bar{\gamma}_1 = .58$. The curve through the saddle singularity x_s as calculated for this case is shown in Figure 14.5, together with the locations of x_0 and x_c.

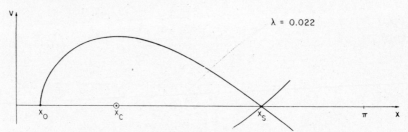

FIG. 14.5. Curve yielding pull-out torque for $\lambda = .022$, $\bar{\gamma}* = 0.8$.

b) $\lambda = .167$, $\bar{\gamma}_* = .8$ with the results: $\gamma_0 = -.026$, $x_0 = -.026$, $\bar{\gamma}_1 = .826$. The curve through the saddle singularity x_s is shown in Figure 14.6. We note in this case that the suddenly applied torque (corresponding to $\bar{\gamma}_1$) is opposite in sign to the initial torque, which means that we have a case in which the *direction* of the torque is suddenly reversed to create the pull-out situation. In Figure 14.6 an integral curve beginning at $v = 5.36$ is indicated: this is a curve corresponding to what Vlasov (cf. Minorsky [31], p. 123) calls a *periodic solution of the second kind* in which v returns to the same value after each complete revolution. In terms of the analogous pendulum problem it means that the pendulum turns indefinitely in the same sense about its support, returning with the same velocity at the end of each full revolution. This is mechanically possible since the external torque can supply just enough energy to the system to make up for losses through the friction force. Vlasov shows that one such motion always exists.

c) $\lambda = .011$, $\bar{\gamma}_* = .4$ with the results: $\gamma_0 = -.84$, $x_0 = -1.00$, $\bar{\gamma}_1 = 1.24$. These results are indicated in Figure 14.7. As we observe, the case is again one like case b) in which the pull-out results from sudden reversal of the torque.

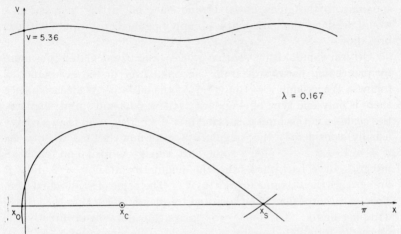

Fig. 14.6. Curve yielding pull-out torque for $\lambda = .167$, $\bar{\gamma}* = 0.8$.

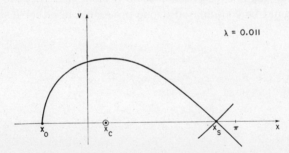

Fig. 14.7. Curve yielding pull-out torque for $\lambda = .011$, $\bar{\gamma}* = 0.4$.

Finally, we remark that still another type of phenomenon in relation to the pull-out problem can be discussed quantitatively using methods like the above. This is the occurrence of a *critical damping factor* (noticed by Lyon and Edgerton [28]) which is defined by the following considerations: Suppose that the initial torque γ_0 is held fixed but that the damping coefficient λ is gradually increased, starting from zero. The critical additional torque corresponding to $\bar{\gamma}_1$ which can be applied without causing instability will also increase, but it can never exceed the value $\bar{\gamma}_1 = 1 - \gamma_0$ at which the total critical torque $\bar{\gamma}_1 + \gamma_0 = \bar{\gamma}_* = 1$ since for any torques such that $\gamma > 1$ there exist no stable equilibrium positions at all. Furthermore it turns out that such an upper limit $\bar{\gamma}_1 = 1 - \gamma_0$ occurs for a finite value λ_* of λ, and it is natural to call λ_* the critical damping factor corresponding to the given initial torque since no further increase in λ would lead to the possibility of applying higher torques without breakdown.

Critical values λ_* for λ with a given γ_0 can be calculated once more by proceeding backwards from the singularity in the x,v-plane, as follows: We assume $\gamma = 1$ in (14.6) at the outset. As a consequence there is only one type of singularity (the saddle and spiral singularities coalesce in this limit case), and it is of a higher order than we have usually considered; the singularities are now located at $n(\pi/2)$, $n = \pm 1, \pm 2, \cdots$. We assume next a value λ_* for λ and follow an integral curve backward from the singularity at $x = \pi/2$ until it first crosses the x-axis again at $x = x_0$. The value of γ_0 obtained from $\sin x_0 = \gamma_0$ corresponds to the initial torque for which λ_* is the critical damping factor. Of course, we have again not solved directly the originally formulated problem, in which γ_0 is given and λ_* is to be calculated, but it is easy to show once more that the same *pairs* of values would be coordinated if one proceeded in either fashion.

CHAPTER IV

Forced Oscillations of Systems with Nonlinear Restoring Force

1. Introduction

The differential equation

$$(1.1) \qquad \ddot{x} + c\dot{x} + f(x) = F \cos \omega t,$$

in which $f(x)$ is nonlinear, occurs in several different kinds of physical problems. The obvious example is the pendulum with an external periodic force applied. The problem of finding the forced oscillation of a single mass subjected to an elastic restoring force leads in general to equation (1.1) if the amplitude of the motion is not kept small. The study of alternating current circuits containing iron core inductances also leads to equation (1.1). The problem of hunting of synchronous electrical machinery is still another example of a physical problem which leads to the same equation.

Since equation (1.1) contains the time t explicitly it cannot be treated by the methods of the preceding chapters, which were based upon a geometrical discussion of the velocity-displacement plane. Although explicit solutions in terms of the elementary functions are not to be expected, the differential equation (1.1) can be treated by various analytic approximation methods, and one of the principal objects of this chapter is to develop some of these analytical methods and to compare and contrast them.

From the theory of differential equations it is known that (1.1) possesses solutions $x(t)$ which are uniquely determined once the values of the displacement and the velocity at the time $t = 0$ are given, i.e., when $x(0)$ and $\dot{x}(0)$, the initial conditions, are prescribed. It is of course clear that solutions of (1.1) other than periodic solutions exist. However, the literature on the subject is almost entirely devoted to the periodic solutions. Apparently the experimenters

81

usually observe the motions to be periodic, at least after some transients have died out, perhaps because the damping forces act in such a way that the motions in a wide variety of cases tend to periodic motions as the time increases.   We concentrate our attention therefore on the study of various types of *periodic solutions* of (1.1) with the object of obtaining in the first place information of a *qualitative* character about the periodic solutions.   Once this has been done, it is relatively easy to see how one should proceed in order to obtain accurate quantitative information.

The approximation methods we use in this chapter can be classed as either perturbation methods or iteration methods.   Furthermore, each of these methods can be applied to our problems in at least two different ways: We may operate directly with the differential equation by using either perturbation or iteration schemes; or, since we assume always that our solutions are periodic, we may assume for them a Fourier series development with undetermined coefficients, and then solve the nonlinear relations which the coefficients must satisfy by applying either iteration or perturbation methods.   One of the principal objects of the present chapter is to illustrate these possibilities in various concrete cases.

We shall treat the *harmonic* oscillations first, i.e. periodic solutions $x(t)$ of (1.1) in which the period is the same as the period $2\pi/\omega$ of the external force $F \cos \omega t$.   Afterwards *subharmonic* oscillations, in which the solution $x(t)$ has as its least period an integral multiple (different from unity) of the period of the external force, will be treated for cases with damping as well as without damping.   There exist also *ultraharmonic* and *ultra-subharmonic* solutions of (1.1) (cf. Chapter I, Section 4 for the definitions of these terms), which could be obtained by the methods used to study the harmonic and subharmonic oscillations, but we shall not treat them here.

Very little generality is lost by choosing for the restoring force $f(x)$ in (1.1) the following cubic in x:

$$(1.2) \qquad f(x) = \alpha x + \beta x^3, \qquad \alpha > 0.$$

The essential qualitative differences in the periodic motions as determined by differences in the character of the spring force are largely due to the distinction between hard ($\beta > 0$) and soft ($\beta < 0$) springs, in the terminology already used in previous chapters.   In the remainder of this chapter we shall usually assume the spring force to be given by (1.2) so that equation (1.1) becomes

$$(1.3) \qquad \ddot{x} + c\dot{x} + (\alpha x + \beta x^3) = F \cos \omega t.$$

Equation (1.3) is often called Duffing's equation since it was Duffing [9] who obtained the first significant results concerning the harmonic solutions of the equation.

2. Duffing's method for the harmonic oscillations without damping

As we have already remarked, explicit solutions of an elementary character are not known for the Duffing equation (1.3). In fact, as we shall see, this comparatively innocent looking differential equation possesses a great variety of periodic solutions alone for which the underlying mathematical theory has been explored to only a slight degree. Still less is known about the nonperiodic solutions. However, as we have already indicated, much that is of interest and value can be achieved by various approximation methods, particularly with regard to the periodic solutions. In this section we shall consider the harmonic solutions, i.e. those periodic solutions which have the same frequency as the impressed force $F \cos \omega t$.

We begin by explaining the iteration method used by Duffing, assuming no damping. The first step is to write the differential equation (1.3) (with $c = 0$) in the form

$$(2.1) \qquad \ddot{x} = -\alpha x - \beta x^3 + F \cos \omega t.$$

We now start with $x_0 = A \cos \omega t$ as first approximation* to the desired solution of frequency ω, and insert it in the right-hand side of (2.1) to obtain

$$(2.2) \qquad \ddot{x}_1 = -(\alpha A + \tfrac{3}{4}\beta A^3 - F) \cos \omega t - \tfrac{1}{4}\beta A^3 \cos 3\omega t$$

as equation for the next approximation x_1, use having been made of the identity

$$(2.3) \qquad \cos^3 \omega t = \tfrac{3}{4} \cos \omega t + \tfrac{1}{4} \cos 3\omega t.$$

* Actually $x_0 = A \cos \omega t$ should be interpreted as the *second* approximation with $x \equiv 0$ as first approximation.

Upon integrating (2.2) twice the result is

$$(2.4) \qquad x_1 = \frac{1}{\omega^2} (\alpha A + \tfrac{3}{4}\beta A^3 - F) \cos \omega t + \tfrac{1}{36} \frac{\beta A^3}{\omega^2} \cos 3\omega t.$$

We have taken the integration constants to be zero in order to ensure that x_1 and the next approximation x_2 be periodic. Before considering what should be done next we observe that an iteration process based on reinserting each successive approximation in the right-hand side of (2.1) in order to obtain the next approximation must require for convergence that the constants α, β, A, and F be sufficiently small. A little later we shall remove this restriction on α and A, but we shall assume that β is small except in Section 11, where the method of Rauscher is treated. The results we obtain are therefore subject to the limitation that the nonlinearity in the spring force is small. We usually also assume that F is small of the same order as β—in other words that the motion takes place in the neighborhood of the free linear vibration of the system—although this latter restriction is not very essential.

So far the iteration method sketched out above is quite straightforward. What to do from now on is not so clear. One might seek to continue the process simply by inserting x_1 in the right-hand side of (2.1) in order to find an x_2, etc., which would be a rather natural procedure. The significance of such a procedure can be explained in a general way in terms of the *response diagram* for the motion. This is a diagram showing the relation between the amplitude of the oscillation (defined, say, as the maximum numerical value of x) and the frequency ω for each value F of the amplitude of the excitation. These curves should lie in the vicinity of the corresponding curves for the *linear forced oscillation* since β is supposed to be small. In Figure 2.1 we indicate the response curves schematically for the linear oscillation (shown dotted) as well as for the nonlinear oscillation (cf. Chapter I, Section 3 for a discussion of the linear response curves). As our remarks above indicate, the iteration scheme now under discussion could yield accurate response curves only for $|A|$ small enough, i.e. near the ω-axis in Figure 2.1. It is therefore clear that there would be difficulties near $\omega = \sqrt{\alpha}$. In addition, this procedure means that the frequency ω is held fixed while the value of the amplitude $|A|$ is left to be determined as a function of it, and this, as we shall see later on, makes it difficult if not impossible to obtain the really essential features of the response curves.

The iteration method introduced by Duffing proceeds along different lines, as follows: The coefficient A_1 of $\cos \omega t$ in (2.4) is taken equal to A, the amplitude of the first approximation, on the ground that A_1 should differ but little from A if $x_0 = A \cos \omega t$ is truly a reasonable first approximation. Also, Duffing argues, such a procedure furnishes the exact result in the linear case ($\beta = 0$) and might hence be expected to yield good results for β small, which is always assumed. This reasoning of Duffing thus leads to the following relation

$$(2.5) \qquad \omega^2 = \alpha + \tfrac{3}{4}\beta A^2 - \frac{F}{A}$$

between the frequency ω and the "amplitude" A of the periodic solution. *The relation (2.5) is basic for the remainder of our discussion;*

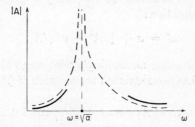

FIG. 2.1. Schematic diagram for nonlinear response.

it has been written purposely so that ω^2 *is given as a function of A* since it is very important to regard A as prescribed and ω as a function of it to be determined.

Before discussing the significance and interpretation of (2.5), it is useful to obtain it again by an iteration procedure which is slightly different from that of Duffing and has the advantage that the spring constant α need not be considered small. This procedure starts with the differential equation

$$(2.6) \qquad \ddot{x} + \omega^2 x = (\omega^2 - \alpha)x - \beta x^3 + \beta F_0 \cos \omega t,$$

which differs from (2.1) only through the addition of the term $\omega^2 x$ to both sides and in the fact that we have put in evidence that $F = \beta F_0$ is small of order β.

It is natural to start the iteration process with the solution of (2.6) for $\omega^2 = \alpha$, $\beta = 0$, i.e., *with the free linear oscillation of frequency* $\omega = \sqrt{\alpha}$. That is, we start with $x_0 = A \cos \omega t$ as first approximation, the "amplitude" A being supposed given.* If we insert x_0 in the right-hand side of (2.6), we find as differential equation for the next approximation x_1 :

(2.7) $\ddot{x}_1 + \omega^2 x_1 = [(-\alpha + \omega^2)A - \frac{3}{4}\beta A^3 + \beta F_0]\cos \omega t - \frac{1}{4}\beta A^3 \cos 3\omega t.$

In obtaining (2.7) we have again made use of the identity (2.3). The equation (2.7) presents the exceptional resonance case since the right-hand side contains a term $P_1 \cos \omega t$ of the same form as would occur in the solution of the homogeneous differential equation. If P_1 were not zero, the solution of (2.7) would contain a term of the type $Ct \sin \omega t$ and $x_1(t)$ could not be periodic. Since we are interested only in the periodic solutions, we must require that P_1, the coefficient of $\cos \omega t$ in (2.7), be zero. Equating this coefficient to zero yields the following relation between the frequency ω and the amplitude A of the first approximation:

(2.8) $\omega^2 = \alpha + \frac{3}{4}\beta A^2 - \beta F_0/A,$

which is exactly the same as Duffing's relation (2.5) with $F = \beta F_0$. Once this relation has been satisfied the solution of (2.7) is

(2.9) $x_1(t) = A_1 \cos \omega t + \dfrac{\beta A^3}{32\omega^2} \cos 3\omega t,$

in which the term $A_1 \cos \omega t$ is the free oscillation with an arbitrary amplitude A_1 *furnished by the homogeneous equation* (as stated earlier we ignore terms in $\sin \omega t$). The second approximation $x_1(t)$ is not fixed until the value of A_1 is prescribed. At this point the decisive step is taken (following Duffing) by choosing for A_1 the value A of the amplitude of the first approximation $x_0 = A \cos \omega t$, and this principle is to be followed in all successive steps in the iteration process. Accordingly we would have, from (2.9):

(2.10) $x_1(t) = A \cos \omega t + \dfrac{\beta A^3}{32(\alpha + \frac{3}{4}\beta A^2 - \beta F_0/A)} \cos 3\omega t,$

in which the value of ω^2 from (2.8) has been used. This procedure means that *the frequency ω is not considered to be prescribed in advance,*

---

* A term $B \sin \omega t$ could also be added, but B would turn out to be zero in the next step of our iteration process. One could, in fact, show that only terms $A_n \cos n\omega t$ with n odd would appear. We have therefore ignored all sine terms in what follows.

but rather to be given by (2.8) *after the value of A has been prescribed.*
This brings with it, however, the necessity to regard the frequency ω
as a function of A, which is somewhat unnatural since it is ω which
figures in the differential equation, and one would ordinarily regard
it as given at the outset. Why this at first sight unnatural seeming
procedure should be the correct one will be made clear a little later.

Before taking up a detailed discussion of the highly important relation (2.8) it is perhaps worth while to consider what would occur in
the next step of the iteration process leading to x_2. Naturally one
would insert the function x_1 from (2.9) in the right side of (2.6).
After conversion of powers and products of cos ωt and cos $3\omega t$ into
sums of terms of the type K cos $n\omega t$, one would obtain for x_2 a differential equation of the form

(2.11) $\quad \ddot{x}_2 + \omega^2 x_2 = P_2 \cos \omega t + Q_2 \cos 3\omega t + R_2 \cos 5\omega t + \cdots ,$

in which P_2, Q_2, etc. would be expressions involving A and ω. In
order to insure the periodicity of the solution $x_2(t)$ it would be necessary to set $P_2 = 0$. This would, of course, yield an improved relation
between ω and A analogous to (2.8). Once P_2 has been set equal to
zero the integration of (2.11) would proceed as before, always with
A cos ωt taken as leading term.

The reasons for our choice of iteration procedure can be readily
seen from a discussion of the relation (2.8). We regard this relation
as furnishing a set of curves in an ω, A-plane with the amplitude
$F = \beta F_0$ of the applied force as parameter. These curves are called
response curves; they reduce to the well known response curves for
linear forced oscillations when $\beta = 0$, as one easily sees. In order
to understand the iteration procedure it was important to consider ω
as depending upon A. However, the response curves are usually
plotted with A as ordinate and we shall follow this practice here.
The quantity A—or, better, its numerical value $| A |$—will also be
referred to on occasion as the "amplitude" of the nonlinear oscillation; actually it is the value of the first Fourier coefficient of the oscillation. In any case, the quantity A, together with ω, characterizes
completely a periodic solution.

In order to obtain the response curves, it is convenient to sketch
first, in an A, ω^2-plane, the curve $\omega^2 = \alpha + 3\beta A^2/4$, and the set of
curves $\omega^2 = -F/A$ for various values of F (with F always taken
positive). For $\beta > 0$, the case of a hard spring force, these curves are

shown in Figure 2.2(a) in which $F_j > F_i$ if $j > i$. The result of adding abscissas in order to obtain the curves given by (2.8) is the set of

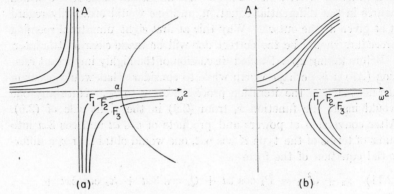

(a)   (b)

FIG. 2.2. Determination of the response curves.

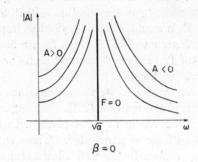

$\beta = 0$

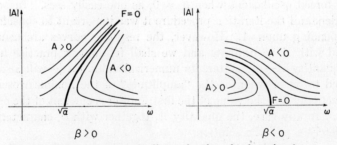

$\beta > 0$ $\beta < 0$

FIG. 2.3. Response curves for linear, hard, and soft spring forces.

curves shown in Figure 2.2(b). In practice it is the custom to plot $|A|$ against ω, as shown in Figure 2.3, where response curves for

$\beta \gtrless 0$ are indicated schematically. The response curve for the free oscillation, corresponding to $F = 0$, is drawn heavier than the others. It is to be noted also that the quantity A is negative on the response curves to the right of the curve for $F = 0$ and positive to the left of it (cf. Figure 2.3), which means that the motion is (to the order of approximation considered) in phase with the external force or 180° out of phase with it according to whether the frequency is greater or less than the frequency of the free oscillation for that particular amplitude. In this respect the behavior of the nonlinear vibration is the same as that of the linear vibration (cf. Chapter I, Section 3). *One sees that the response curves in the nonlinear cases could be thought of as arising from those for the linear case by bending the latter to the right for a hard spring and to the left for a soft spring.*

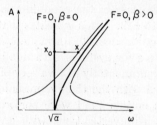

Fig. 2.4. Diagram indicating the character of the approximation method.

We can now see also why the quantity A should be prescribed while the frequency ω should be considered as a function of A: *For certain values of ω there are three corresponding values of A; hence the first iteration process sketched above, which at the outset seems a rather natural one, could not possibly yield all branches of the response curves.* We note further that we began our iterations with a free linear oscillation, i.e., we started with $\omega = \sqrt{\alpha}$ and $F = 0$ and $\beta = 0$. By iteration we then passed to a neighboring solution, as indicated in Figure 2.4, and since A is arbitrary in the linearized problem *its value must be assumed to begin with.* For these reasons the method of Duffing yields the significant results in our present problem which earlier workers—including even Rayleigh—failed to obtain.

For $F = 0$ we obtain from the preceding discussion the free oscillations. In Chapter III we have already seen that the free oscillation in this case can be treated by explicit integration and the amplitude of the free oscillation can be found as a function of the frequency.

In the case of the free oscillation the response relation (2.8) becomes

$$(2.12) \qquad\qquad \omega^2 = \alpha + \tfrac{3}{4}\beta A^2,$$

and it gives the exact result $\omega^2 = \alpha$ in the linear case ($\beta = 0$). One might expect it to be accurate, in addition, for not too large values of βA^2. In order to obtain an indication of the accuracy which may be expected from (2.12), and so perhaps also from (2.8), consider the case of the pendulum, for which $\beta = -\alpha/6$. In this case (2.12) is

$$(2.13) \qquad\qquad \omega^2 = \alpha(1 - \tfrac{1}{8}A^2)$$

and this result can easily be seen to coincide with the first two terms of the exact power series for ω^2 as a function of the amplitude A of the pendulum. (This could be obtained from equation (4.15) of Chapter II by identifying A with a and observing that $\omega = 2\pi/T$.) Formula (2.13) for the case of the pendulum is quite accurate even for amplitudes A up to $\pi/2$. This indicates that the second approximation $x_1(t)$ may be quite accurate for moderately large values of A.

The response curves for systems with nonlinear restoring forces have been checked experimentally in a variety of different cases, with results in good accord with the theory presented above. Probably the first experiments of this kind were performed by Martienssen [29], who carried out his investigations with an electrical apparatus involving a condenser and an inductance with an iron core; he was aware of the influence of the nonlinearity. Duffing [9] checked the results of his theory by experiments using a pendulum, and found good numerical agreement.

3. The effect of viscous damping on the harmonic solutions

In this section we give a method for approximating the harmonic solutions of Duffing's equation when the viscous damping term $c\dot{x}$, $c > 0$, is included.

If damping is neglected we have seen that there is an oscillation $x = A \cos \omega t$ either in phase with the impressed force $F \cos \omega t$ or 180° out of phase with it. If damping is present, however, the displacement and the impressed force can be expected to be out of phase, just as in the case of the corresponding linear problem. In order to take into account this difference in phase, the impressed force could be prescribed and the phase of the solution left to be determined, but it is more convenient to fix the phase of the solution and leave the phase of the impressed force as a quantity to be determined. Hence the differential equation is taken in the form

$$(3.1) \qquad \ddot{x} + c\dot{x} + (\alpha x + \beta x^3) = H \cos \omega t - G \sin \omega t$$

in which the amplitude $F = \sqrt{H^2 + G^2}$ of the impressed force is considered as fixed, but the ratio H/G is left to be determined. We assume also that c, G, and H are all small of order β. Guided by the experience gained in the preceding case we now proceed as follows: As first approximation to the solution we take $x = A \cos \omega t$, with A regarded as fixed and ω as a quantity to be determined. Insertion of this in (3.1) then leads to the following two equations, when (2.3) is used and the term in $\cos 3\omega t$ is neglected:

$$(3.2) \qquad (\alpha - \omega^2)A + \tfrac{3}{4}\beta A^3 = H$$

$$(3.3) \qquad A c \omega = G.$$

These equations, incidentally, verify a conclusion reached in Chapter III: *No periodic motion except the state of rest can exist if there is damping but no external force, for $H = G = 0$ implies $A = 0$ if $c \neq 0$.*

The relation between the amplitude of the impressed force and the quantities A and ω is readily obtained by squaring (3.2) and (3.3) and adding. The result is

$$(3.4) \qquad [(\alpha - \omega^2)A + \tfrac{3}{4}\beta A^3]^2 + c^2 A^2 \omega^2 = H^2 + G^2 = F^2.$$

It is convenient to write (3.4) in the form

$$(3.5) \qquad S^2(\omega, A) + c^2 A^2 \omega^2 = F^2,$$

in which

$$(3.6) \qquad S(\omega, A) = (\alpha - \omega^2)A + \tfrac{3}{4}\beta A^3.$$

We observe that $S(\omega, A) = F$ is the relation which yields the response curves shown in Figure 2.3 for the case in which there is no damping. Since c is assumed to be small of order β it is to be expected that the curves for (3.5) differ only slightly from those of Figure 2.3, and that *they are rounded off in the vicinity of the curve for $F = 0$.* The latter expectation is certainly justifiable when β is sufficiently small, since in this case (3.5) represents curves only slightly different from the linear response curves of Figure 3.1, Chapter I. The response curves which result from (3.5) are indicated schematically in Figure 3.1, both for a hard spring ($\beta > 0$) and a soft spring ($\beta < 0$) as well as for a linear spring ($\beta = 0$). Again we observe that these curves could be considered as arising from the linear response curves by bending the latter to the right or left, depending on the type of spring. In Figure 3.1 we have drawn dotted curves through the points of the

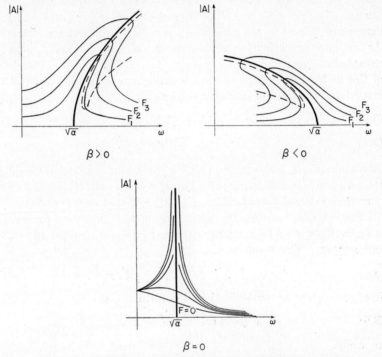

FIG. 3.1. Response curves when viscous damping is present.

response curves having vertical tangents; as one sees, these dotted curves form the outline of the region in which the response curves turn over on themselves. The significance of this region from the physical point of view will be explained shortly. We may, however, note here (cf. Figure 3.1) that the turning over of the response curves takes place only when the amplitude F of the excitation is larger than a certain value.

In order to carry out a complete discussion of the response curves it is thus obviously useful to determine the loci of the vertical tangencies, and in fact also of the horizontal tangencies. The locus of the points of contact of the *horizontal tangents* of (3.5) can easily be found by differentiating (3.5) implicitly with respect to ω and setting $dA/d\omega$ equal to zero. The equation of the locus is therefore given by $S(\omega, A) = c^2A/2$ or

$$(3.7) \qquad \omega^2 - \tfrac{3}{4}\beta A^2 = \alpha - c^2/2.$$

The last equation is the same as (2.12) except for the term $-c^2/2$. The curve therefore is a conic of the same kind as the response curve for a free oscillation and tends to it from the left side as c tends to zero. There is only one horizontal tangent on each of the response

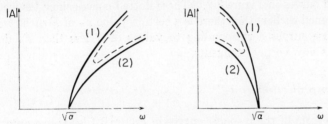

FIG. 3.2. Loci of the vertical tangencies on the response curves.

curves furnished by (3.5). It can be shown easily that A has a maximum at the points with a horizonal tangent. The equation of the locus of the *vertical tangents* can be found in a similar way by differentiating (3.5) implicitly with respect to A and setting $d\omega/dA$ equal to zero. The result is

$$(3.8) \qquad (\alpha - \omega^2 + \tfrac{3}{4}\beta A^2)(\alpha - \omega^2 + \tfrac{9}{4}\beta A^2) + c^2\omega^2 = 0.$$

If there is no damping, that is if $c = 0$, equation (3.8) yields the following pair of equations:

$$(1) \qquad \omega^2 - \tfrac{3}{4}\beta A^2 = \alpha,$$

$$(2) \qquad \omega^2 - \tfrac{9}{4}\beta A^2 = \alpha.$$

The curves corresponding to these equations are shown in Figure 3.2. The first of these is, as one easily verifies, the response curve for the free undamped oscillation; the second is the locus of the vertical tangents of the curves in Figure 2.3. Hence if c is small one can expect the relation (3.8), which characterizes the vertical tangencies for

$c \neq 0$, to furnish a curve near these two; in particular, there should be one branch near the response curve for the free oscillation, and another near the points where the curves for $c = 0$ turn over on themselves. In other words, the curves must appear as they are shown in Figure 3.1. The important curves represented by equation (3.8) which yield the vertical tangencies can be conveniently studied if one first makes the substitutions $\omega^2 = x$, $A^2 = y$. In the x, y-plane (3.8) represents a hyperbola. For $\beta > 0$, one branch of the hyperbola lies entirely in the first quadrant, the other branch lies below the x-axis; for $\beta < 0$, one branch of the hyperbola lies in both the first and second quadrants, the other branch lies below the x-axis. It is then easy to see that in the ω, A-plane (3.8) gives curves of the forms shown by the dotted curves in Figure 3.2. These dotted curves show (as we have mentioned earlier) that there is a certain value F_* of F such that the response curves corresponding to values of F less than F_* do not possess any vertical tangents.

4. Jump phenomena

The form of the response curves in Figure 3.1 leads to a number of conclusions of importance for the physical phenomena. In Figure 4.1 representative curves for $\beta > 0$ and $\beta < 0$ are indicated. Let us imagine an experiment performed in which the amplitude F of the external force is held constant, while its frequency is slowly varied and the amplitude A of the resulting harmonic oscillation is observed. Consider first the case of a hard spring force, $\beta > 0$, and suppose that ω is rather large at the beginning of our experiment, i.e., we start at point 1 on the curve. As ω is decreased A slowly increases through the point 2 until the point 3 is reached. Since F is held constant, a further decrease in ω would require a jump from point 3 to point 4 with an accompanying increase in the amplitude A, after which A decreases with ω. Upon performing the experiment in the other direction, i.e., starting at the point 5 and increasing ω, the amplitude follows the 5-4-6 portion of the curve, then jumps to point 2 and afterwards slowly decreases. The circumstances are quite similar with a soft spring force, but the jumps in amplitude take place in the reverse direction. It would not have been necessary, we observe, to consider the influence of damping in order to conclude that a jump from point 3 to point 4 should take place (for $\beta > 0$) on decreasing ω, but the jump from point 6 to point 2 on increasing ω would be inexplicable on the basis of the curves of Figure 2.3, which should be taken when there is no damping.

Instead of performing an experiment in which the force amplitude F is held constant while the frequency ω is varied, one might vary F while holding ω fixed. In fact, the latter experiment would probably be simpler to carry out. For example, it might be done by using an electrical apparatus consisting of an iron core inductance (which is the origin of the nonlinearity in the system) in series with a condenser

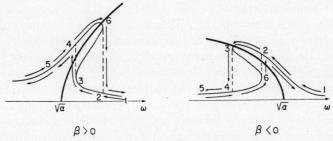

Fig. 4.1. Jump, or hysteresis, phenomena.

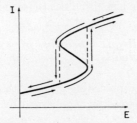

Fig. 4.2. Hysteresis in a nonlinear electrical system.

and an alternating current generator. The frequency can be held constant simply by holding the speed of the generator constant while the voltage E (which is proportional to the force amplitude F) can be varied by changing the field current of the generator. When the voltage E is gradually increased, it is found that the current I rises gradually to a certain value and then increases suddenly to a much larger value after which a further increase in E causes only a gradual rise in the current. If, then, one attempts to reverse the process, it is found that the same curve is not retraced completely, but that the jump from the higher to the lower current takes place at a lower voltage than before, as indicated in Figure 4.2. An analysis of the relation (3.4) between A, ω, and F which results when damping is taken into account leads to similar conclusions when the amplitude of the response is considered as a function of the force amplitude F with ω held constant. These statements can be easily verified either by referring directly to equation (3.4) or to the curves shown in Figure 3.1 and considering the manner of variation of A with F along a line $\omega = $ constant.

Our discussion of the jump phenomena leads us to suspect that the harmonic oscillations corresponding to points on the response curves between the vertical tangencies—in other words, the points lying in the regions bounded by the dotted curves of Figure 3.1—are in some sense unstable. This is an important matter which will be discussed briefly in the last section of this chapter and in much greater detail in Chapter VI from the mathematical point of view. We observe here only that the jump phenomena have often been observed experimentally, for example by Martienssen [29] in an electrical system and Duffing [9] in a mechanical system. The effect of viscous damping on the response curves has been discussed by Appleton [2], who also observed and explained the jump phenomena he encountered in working with a certain galvanometer which behaved like the systems under discussion here. Appleton also discusses the stability of the oscillations.

5. Hunting and pull-out torques of synchronous motors under oscillatory loads

We have seen in the preceding chapter that the unsteady operation of the synchronous motor is governed by the equation

$$(5.1) \qquad \ddot{x} + c\dot{x} + k \sin x = L.$$

In the present section we discuss in a qualitative way a few problems of interest in practice when the external torque L is not a constant (as we assumed in the preceding chapter) but a periodic function of the time, as follows:

$$(5.2) \qquad L = L_0 + L_1 \cos \omega t.$$

In other words, the external torque is assumed to consist of a periodic part superimposed upon a constant torque.

The case in which $L_0 = 0$, i.e. in which the load is entirely oscillatory, can be discussed immediately on the basis of the results of the preceding sections. We know that oscillations of frequency ω can occur with a definite amplitude; this condition of operation of the synchronous motor is called hunting. Hunting of a motor may be very undesirable, particularly if the amplitude of the oscillation should become too large: undesirable surges in the electrical system supplying power to the motor would occur, or the motor could be thrown out of step through skipping a pole just as in the cases discussed in the preceding chapter. The following problem therefore has practical significance: Determine the range of frequencies ω and amplitudes L_1 of the disturbing torque such that the amplitude $|A|$ of the response will never exceed a given value $|A|_{max}$. This problem can be solved (for the special case $L_0 = 0$, i.e. the case of zero average load) by making use of the appropriate response curves

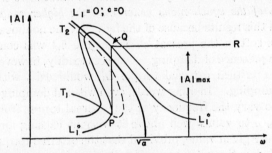

FIG. 5.1. Limiting the amplitude arising from the hunting of a synchronous motor or generator.

determined by the methods discussed above. In Figure 5.1 we indicate schematically how one can obtain the region containing all points for which the frequency ω and corresponding values of the amplitude L_1 of the external force lead to amplitudes never greater than $|A|_{max}$. On account of the jump phenomena discussed in the preceding section it is not sufficient simply to draw a line $|A| = |A|_{max}$ in the $|A|$, ω-plane and take the region under this line as the desired region. Instead, the region in question lies below the heavy broken curve T_1PQR, which follows the lower branch of the locus of vertical tangents of the response curves from T_1 to a certain point P, rises vertically to Q, and then follows the horizontal line $|A| = |A|_{max}$. In this way those unstable oscillations with $|A| < |A|_{max}$, from which jumps to stable oscillations with $|A| > |A|_{max}$ could occur, are ruled out. The point P is fixed by finding the "critical response curve" associated with a value L_1^* of the external torque such that the vertical tangent drawn on it from the point P intersects the upper branch of *the same response curve* at Q on the line $|A| = |A|_{max}$. Points to the left of PQ and above T_1P lie on "unstable" branches of response curves whose upper stable branches correspond to values of $|A|$ larger than $|A|_{max}$, while points to the right of PQ lead to values of $|A|$ less than $|A|_{max}$. The critical value L_1^* of the external torque thus has the following property: If the external torque L_1 is kept below this value the amplitude $|A|$ of the motion will be less than $|A|_{max}$ for all frequencies, but this will not hold good for all frequencies if $L_1 > L_1^*$.

One sees from Figure 5.1 that *the range of frequencies a little below the frequency* $\sqrt{\alpha}$ *of the linear free oscillation is much more critical for the operation of the synchronous motor than the higher range of frequencies*, and this results because of the fact that the nonlinear restoring torque is soft in the present case. Figure 5.1 was constructed assuming the presence of damping. One sees readily, however, that a non-vanishing critical external torque L_1^* would exist without the presence of damping. In other words, even without damping it would be possible to vary the frequency of the external torque from values well below $\sqrt{\alpha}$ to values well above it without causing amplitudes higher than any given value, provided that the external torque amplitude is kept below the critical value.† This is, of course, not possible with a linear restoring force.

If the constant part L_0 of the external torque is not zero, the problem of hunting of the synchronous motor can be attacked in much the same way as for $L_0 = 0$. A special case of this kind has been treated by Duffing.

6. The perturbation method

One of the commonest methods for treating nonlinear problems in mechanics is the perturbation method, which consists in developing the desired quantities in powers of some parameter which can be considered small, and determining the coefficients of the developments stepwise, usually by solving a sequence of linear problems. The

---

† Naturally, this statement (as well as other earlier statements) is made only on the assumption that it is the harmonic oscillation studied here which is actually excited.

method has the advantage that it is relatively foolproof, in the sense that it can often be applied fairly safely even in the absence of fore-knowledge regarding the general character of the solution.* The method of perturbations is also very useful in settling important theoretical questions of a purely mathematical character; for example, the existence of the various types of periodic motions discussed in this chapter can probably be established most readily by proving that the appropriate perturbation series converge. In Appendix I to this book we take up such questions in some detail. However, the perturbation method has the disadvantage that it is often rather cumbersome for actual computations, particularly if more than one or two terms in the perturbation series are desired. Consequently, it is often advantageous to begin the attack on a new problem by the perturbation method in order to gain a first insight into the character of its solution, but to abandon this attack eventually in favor of other approaches once sufficient knowledge about the behavior of the solutions has been gained. For this reason, as well as for reasons indicated at the beginning of this chapter, we treat the Duffing problem in detail once more by the perturbation method as applied directly to the differential equation.

In this section we illustrate the use of the perturbation method (in a form first used by Lindstedt [26], apparently, in connection with problems in astronomy) to obtain the harmonic solutions of

$$(6.1) \qquad \ddot{x} + (\alpha x + \beta x^3) = F \cos \omega t.$$

As we have stated, the method consists in developing the desired solution $x(t)$ in a power series with respect to a small parameter ϵ, the coefficients in the series being functions of t. We write, therefore,

$$(6.2) \qquad x = x_0 + \epsilon x_1 + \epsilon^2 x_2 + \cdots,$$

the x_i being functions of t. A periodic solution $x(t)$ is desired which has the same frequency as $F \cos \omega t$. It would be natural to regard the amplitude of $x(t)$ as a quantity to be determined for any given frequency ω. However, our previous experience has taught us that the amplitude of the vibration rather than its frequency should be prescribed. In order to avoid working with functions of unknown

---

* The problems under consideration here, however, are not altogether of this character. One must gain some advance insight about the character of the solutions in order to fix the details of the procedure.

period it is of advantage to introduce a new independent variable θ replacing t through the relation $\theta = \omega t$. The equation (6.1) becomes

$$(6.3) \qquad \omega^2 \frac{d^2 x}{d\theta^2} + (\alpha x + \beta x^3) = F \cos \theta$$

We now require that $x(\theta)$ satisfy the following conditions:

$$\text{a)} \qquad x(\theta + 2\pi) = x(\theta)$$

$$(6.4) \qquad \text{b)} \qquad x(0) = A$$

$$\text{c)} \qquad x'(0) = 0.$$

The prime on a quantity means differentiation with respect to θ here and in what follows. The condition a) states that $x(\theta)$ is to be of period 2π, while b) and c) fix, roughly speaking, the amplitude and phase of the vibration. It should be noted that A is the maximum of $x(t)$ here rather than the first Fourier coefficient, and hence is not quite the same as the quantity A in the preceding sections. The value of ω will depend upon A, as we know.

The parameter ϵ is arbitrary to a certain degree in any perturbation procedure. It is, however, natural in this case to choose $\epsilon = \beta$, so that the perturbation series (6.2) may be considered as a development in the neighborhood of the solution of the linearized vibration problem. In addition to x it is also necessary to develop the quantity ω with respect to β. If the series

$$(6.5) \qquad x(\theta) = x_0(\theta) + \beta x_1(\theta) + \beta^2 x_2(\theta) + \cdots ,$$

$$(6.6) \qquad \omega = \omega_0 + \beta \omega_1 + \beta^2 \omega_2 + \cdots$$

are inserted in (6.3) we obtain a power series in β which must vanish identically in β; hence the coefficients of the successive powers of β must vanish. The coefficients are second order linear differential equations in the $x_i(\theta)$, which involve also the constants ω_i. To determine the x_i and the ω_i we have the conditions (6.4) which lead to the new conditions

$$\text{a)} \qquad x_i(\theta + 2\pi) = x_i(\theta)$$

$$(6.7) \qquad \text{b)} \qquad x_0(0) = A, \qquad x_i(0) = 0$$

$$\text{c)} \qquad x_0'(0) = 0, \qquad x_i'(0) = 0.$$

The condition a) will serve to determine the constants ω_i in (6.6).

It is more convenient, though not strictly necessary, to assume that the amplitude F of the applied force is also small with β; we assume, therefore:

$$(6.8) \qquad\qquad F = \beta F_0.$$

This means that our development is one in the neighborhood of the linear *free* vibration, just as it was in the treatment of the same problem by the iteration method.

The result of inserting (6.5), (6.6), and (6.8) in (6.3) is

$$(6.9) \qquad (\omega_0^2 + 2\beta\omega_0\omega_1 + \cdots)(x_0'' + \beta x_1'' + \cdots)$$
$$+ \alpha(x_0 + \beta x_1 + \beta^2 x_2 + \cdots)$$
$$+ \beta(x_0^3 + 3x_0^2 x_1\beta + \cdots) = \beta F_0 \cos\theta.$$

By x_i'' is meant, of course, $d^2 x_i/d\theta^2$.

The term of zero order in β yields

$$(6.10) \qquad\qquad \omega_0^2 x_0'' + \alpha x_0 = 0,$$

the general solution of which is

$$(6.11) \qquad x_0 = A_0 \cos \frac{\sqrt{\alpha}}{\omega_0}\theta + B_0 \sin \frac{\sqrt{\alpha}}{\omega_0}\theta.$$

The conditions (6.7) lead at once to

$$(6.12) \qquad \begin{array}{ll} \text{a)} & \omega_0 = \sqrt{\alpha} \\ \text{b)} & A_0 = A \\ \text{c)} & B_0 = 0. \end{array}$$

Hence we have determined

$$(6.13) \qquad\qquad x_0 = A \cos\theta$$

and

$$(6.14) \qquad\qquad \omega_0 = \sqrt{\alpha},$$

i.e., the zero order terms in the perturbation series (6.5) and (6.6).

We continue the process by taking the first order term in (6.9). This leads to a differential equation for $x_1(\theta)$:

$$(6.15) \qquad \omega_0^2 x_1'' + \alpha x_1 = -2\omega_0\omega_1 x_0'' - x_0^3 + F_0 \cos\theta.$$

Insertion of x_0 from (6.13) yields

$$(6.16) \quad \omega_0^2 x_1'' + \alpha x_1 = (2\omega_0\omega_1 A - \tfrac{3}{4}A^3 + F_0)\cos\theta - \tfrac{1}{4}A^3\cos 3\theta.$$

The periodicity condition for x_1 requires that the coefficient of $\cos\theta$ be zero, since otherwise a term $\theta\sin\theta$ would arise in the general solution of (6.16), which presents the exceptional resonance case. Hence we set $2\omega_0\omega_1 A - 3A^3/4 + F_0 = 0$, or

$$(6.17) \qquad \omega_1 = \frac{1}{2\sqrt{\alpha}}\left(\frac{3}{4}A^2 - \frac{F_0}{A}\right),$$

which fixes the quantity ω_1 in (6.6). We note that the first two terms in (6.6) yield, in view of (6.12a) and (6.17), the same relation (within terms of order less than β^2) between ω and A as was found previously (cf. equation (2.8)). From $\omega^2 = \omega_0^2 + 2\omega_0\omega_1\beta + \cdots$, we obtain in fact

$$\omega^2 = \alpha + \beta\left(\frac{3}{4}A^2 - \frac{F_0}{A}\right) + \cdots$$

$$= \alpha + \tfrac{3}{4}\beta A^2 - \frac{\beta F_0}{A} + \cdots$$

where the dots refer to terms of order β^2 and higher. The general solution of (6.16) may now be written

$$(6.18) \quad x_1 = A_1\cos\frac{\sqrt{\alpha}}{\omega_0}\theta + B_1\sin\frac{\sqrt{\alpha}}{\omega_0}\theta - \frac{A^3}{4(\alpha - 9\omega_0^2)}\cos 3\theta$$

$$= A_1\cos\theta + B_1\sin\theta + \frac{A^3}{32\alpha}\cos 3\theta$$

upon setting $\omega_0 = \sqrt{\alpha}$. The conditions (6.7b, c) require $A_1 = -A^3/32\alpha$ and $B_1 = 0$. Hence we have finally

$$(6.19) \qquad x_1 = \frac{A^3}{32\alpha}(-\cos\theta + \cos 3\theta).$$

The approximation $x = x_0 + \beta x_1$ can now be seen to coincide with the second approximation (2.10) furnished by the iteration procedure, again within terms of order β^2. Our solution, up to terms of first order in β, is

$$(6.20) \qquad x = A\cos\theta + \beta\frac{A^3}{32\alpha}(-\cos\theta + \cos 3\theta) + \cdots,$$

and

(6.21) $\qquad \omega = \sqrt{\alpha} + \beta \, \dfrac{1}{2\sqrt{\alpha}} \left(\dfrac{3}{4} A^2 - \dfrac{F_0}{A} \right) + \cdots .$

The method of procedure from this point on should be clear.

7. *Subharmonic response*

Up to now we have considered only the harmonic solutions of the Duffing equation, that is, solutions for which the frequency is the same as that of the external force $F \cos \omega t$. Permanent oscillations whose frequency is a fraction $\frac{1}{2}$, $\frac{1}{3}$, $\cdots$, $1/n$, of that of the applied force can, however, occur in nonlinear systems, in particular in our case of the Duffing equation. To this phenomenon the term subharmonic response is usually applied, though the term frequency demultiplication is also used and is perhaps a better one. (For literature on this subject see the papers by Baker [3], Krylov and Bogoliuboff [21], v. Kármán [20], and Friedrichs and Stoker [13].)

The fact that subharmonic oscillations occur in systems with nonlinear restoring forces can hardly be denied since they have been often observed (cf. the paper of Ludeke [27], for example). But it is not an entirely simple matter to give a plausible physical explanation for their occurrence. Let us recall the behavior of linear systems. If the frequency of the free oscillation of a linear system is ω/n (n an integer, say) then a periodic external force of frequency ω can excite the free oscillation in addition to the forced oscillation of frequency ω. But since some damping is always present in a physical system, the free oscillation is damped out so that the eventual "steady state" consists solely of the oscillation of frequency ω. Why should the situation be different in a nonlinear system? The explanation usually offered is as follows: Any free oscillation of a nonlinear system contains the higher harmonics in profusion, and hence it is possible that an external force with a frequency the same as one of these might be able to excite and sustain the harmonic of lowest frequency. Of course that this actually should occur probably requires that the damping be not too great and that proper precautions of various kinds be taken.

We shall not attempt to present a solution of the problem of subharmonic response for the Duffing equation in all generality. Rather, we shall treat only one special case, i.e. the subharmonic oscillation of order $\frac{1}{3}$. Also, in the development which immediately follows, damping will be neglected, but will be taken into account in the next section. We are interested here, as in our previous discussions, in qualitative rather than accurate quantitative results. We are also interested here, as otherwise in this book, in discussing various different methods of solution which can be made useful in treating nonlinear vibration problems. We have therefore chosen to treat the problem of the present section by operating directly with the Fourier series for the solution. Of the two possibilities for calculating the coefficients of the series—that is, either the perturbation or the iteration method—we have chosen the iteration method.

There is some advantage in introducing, as in the preceding section, the variable $\theta = \omega t$ as new independent variable. The differential equation is

$$(7.1) \qquad \omega^2 x'' + (\alpha x + \beta x^3) = F \cos \theta,$$

in which it is to be remembered that ω represents the frequency of the applied force $F \cos \omega t$. Our object is to find a periodic solution with frequency $\omega/3$, or in terms of the variable θ of frequency $\frac{1}{3}$. It follows that the subharmonic solution can be developed in a Fourier series of the form

$$x = \sum_{n=1}^{\infty} a_n \cos \frac{n\theta}{3} + b_n \sin \frac{n\theta}{3}.$$

Our object is to determine the coefficients in the series approximately, on the assumption that β is small. We remark first of all that all sine terms and also all cosine terms in even multiples of $\theta/3$ in the Fourier series turn out to be zero in the course of the calculation; for the sake of brevity we therefore omit these terms at the outset and write our solution in the form:

$$(7.2) \qquad x = A_{1/3} \cos \frac{\theta}{3} + A_1 \cos \theta + A_{5/3} \cos \frac{5\theta}{3} + \cdots.$$

Substitution of (7.2) in (7.1) and use of some of the following identities, in which the dots refer to terms involving higher multiples of θ:

$$\cos^3 \frac{\theta}{3} = \tfrac{3}{4} \cos \frac{\theta}{3} + \tfrac{1}{4} \cos \theta$$

$$\cos^2 \frac{\theta}{3} \cos \theta = \tfrac{1}{4} \cos \frac{\theta}{3} + \tfrac{1}{2} \cos \theta + \cdots$$

(7.3)
$$\cos \frac{\theta}{3} \cos^2 \theta = \tfrac{1}{2} \cos \frac{\theta}{3} + \cdots$$

$$\cos^3 \theta = \tfrac{3}{4} \cos \theta + \cdots$$

$$\cos^2 \frac{\theta}{3} \sin \theta = \tfrac{1}{4} \sin \frac{\theta}{3} + \tfrac{1}{2} \sin \theta$$

$$\cos \frac{\theta}{3} \sin^2 \theta = \tfrac{1}{2} \cos \frac{\theta}{3} + \cdots$$

leads to the relations

$$(7.4) \qquad \left(\alpha - \frac{\omega^2}{9} \right) A_{1/3} + \tfrac{3}{4}\beta(A_{1/3}^3 + A_{1/3}^2 A_1 + 2A_{1/3}A_1^2) = 0,$$

$$(7.5) \qquad (\alpha - \omega^2)A_1 + \tfrac{1}{4}\beta(A_{1/3}^3 + 6A_{1/3}^2 A_1 + 3A_1^3) = F.$$

Equations (7.4) and (7.5) take the place of equation (2.8) which was fundamental for the "harmonic" case. Only the two lowest harmonics have been considered.

Again we are faced with a problem of procedure, but our previous experience offers a guide. We set $\beta = 0$ (i.e., we consider the linear case) and observe from (7.4) that $A_{1/3}$ must be taken zero unless $\omega = 3\sqrt{\alpha}$, and no subharmonic oscillation would be obtained. If, however, $\omega = 3\sqrt{\alpha}$ then $A_{1/3}$ can be taken arbitrarily, while $A_1 = -F/8\alpha$ is determined from (7.5) for a given value of F when $\beta = 0$. The term $A_{1/3} \cos (\theta/3)$ represents evidently the free oscillation of arbitrary amplitude which may be superimposed on the forced vibration $(-F/8\alpha) \cos \theta$ in the linear case. Hence we should prescribe $A_{1/3}$ for $\beta \neq 0$ and then hold it fixed. The quantity F should also be prescribed, but the quantities ω and A_1, which were fixed for $\beta = 0$, should now be considered as functions of $A_{1/3}$ and F. It should be noted that we begin our approximation in the present case with the linear *forced* oscillation rather than with the linear free oscillation, as in the preceding case.

With this in mind we rewrite equations (7.4) and (7.5) in the following form *in order to perform iterations conveniently*. After dividing equation (7.4) by $A_{1/3} \neq 0$, we rewrite equations (7.4) and (7.5) as follows:

(7.6)
$$\begin{cases} \omega^2 = 9\alpha + \tfrac{27}{4}\beta(A_{1/3}^2 + A_{1/3}A_1 + 2A_1^2), \\ -8\alpha A_1 = F + (\omega^2 - 9\alpha)A_1 - \tfrac{1}{4}\beta(A_{1/3}^3 + 6A_{1/3}^2 A_1 + 3A_1^3), \end{cases}$$

or

$$(7.7) \quad -8\alpha A_1 = F - \tfrac{1}{4}\beta(A_{1/3}^3 - 21A_{1/3}^2 A_1 - 27A_{1/3}A_1^2 - 51A_1^3),$$

the last equation resulting from the elimination of ω^2.

We begin the iterations with the values for $\beta = 0$, i.e., with $A_{1/3}$ prescribed, $\omega = 3\sqrt{\alpha}$, and $A_1 = -F/8\alpha = A$. The next step yields

$$(7.8) \qquad \omega^2 = 9\alpha + \tfrac{27}{4}\beta(A_{1/3}^2 + A_{1/3}A + 2A^2),$$

$$(7.9) \quad A_1 = A + \tfrac{1}{32}\frac{\beta}{\alpha}(A_{1/3}^3 - 21A_{1/3}^2 A - 27A_{1/3}A^2 - 51A^3),$$

as one can readily verify.

Equation (7.8) is readily discussed. It represents an ellipse or a hyperbola in an ω, $A_{1/3}$-plane, depending on the sign of β. Also ω^2 has a minimum when $\beta > 0$ and a maximum when $\beta < 0$ for $A_{1/3} = -A/2$ and ω^2 has as value there

$$(7.10) \qquad\qquad \omega^2 = 9(\alpha + \tfrac{21}{16}\beta A^2).$$

Thus the subharmonic vibration exists only for

$$(7.11) \qquad \omega \begin{array}{c} < \\ > \end{array} 3\sqrt{\alpha + \tfrac{21}{16}\beta A^2} \quad \text{when} \quad \begin{array}{c} \beta < 0 \\ \beta > 0. \end{array}$$

If $\beta \neq 0$, we may conclude that *no subharmonic vibration with $\omega = 3\sqrt{\alpha}$ can exist*, i.e., no subharmonic response with exactly the frequency of the free oscillation of the linearized equation can exist. Some authors describe subharmonic response as an oscillation with exactly the frequency of the free oscillation of the linearized equation excited by a force with n times this frequency. In the case of the Duffing equation, at least, such an oscillation cannot occur.

Interesting conclusions can be drawn from the relation (7.9), which determines the second Fourier coefficient of the subharmonic vibration as a function of $A_{1/3}$. For $A_{1/3} = 0$ relation (7.4) is satisfied identically while relation (7.5) reduces to relation (2.8) for the harmonic case. Hence *the subharmonic vibration results through bifurcation from the harmonic vibration.* This takes place for

$$(7.12) \qquad\qquad A_1 = A - \tfrac{5}{3}\tfrac{1}{2}\beta A^3,$$

with $A = -F/8\alpha$.

Figure 7.1 indicates the nature of the curves for $A_{1/3}$ and $|A_1 + A_{1/3}|$ for both a hard and a soft restoring force. In the lower

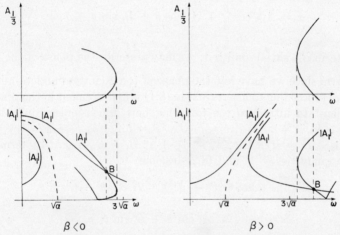

FIG. 7.1. Response curves for a subharmonic oscillation.

two graphs we have indicated the response curves for $|A_1|$, the amplitude of the harmonic oscillations, as well as curves for the "amplitude" $|A_1|$ of the "harmonic" component of the oscillation. The point B is the bifurcation point; the subharmonic and the harmonic oscillations corresponding to it are identical. The intersection point C on the curve for $\beta > 0$ is not a bifurcation point; the harmonic and subharmonic oscillations corresponding to it are different.

8. Subharmonics with damping

We have already remarked in the first chapter that subharmonics in linear systems with damping cannot occur. In nonlinear systems, however, subharmonics can occur even in the presence of viscous damping, as we shall see in the present case of the Duffing equation. As in earlier sections, we introduce $\theta = \omega t$ as independent variable and write the differential equation in the form

$$(8.1) \qquad \omega^2 x'' + \omega c x' + (\alpha x + \beta x^3) = H \cos \theta - G \sin \theta,$$

and assume that the amplitude $F = \sqrt{H^2 + G^2}$ of the excitation is prescribed, but the ratio H/G is not fixed in advance. To determine the subharmonic oscillation of order $\frac{1}{3}$ we start, as in the preceding section, with the Fourier series

$$(8.2) \qquad x = A_{1/3} \cos \frac{\theta}{3} + A_1 \cos \theta + B_1 \sin \theta + \cdots.$$

The term $B_{1/3} \sin \dfrac{\theta}{3}$ is omitted: we may prescribe the phase of the lowest term since we have left the phase of the excitation undetermined. Before proceeding to insert (8.2) in (8.1) it is convenient in the present problem* to introduce the following new dimensionless quantities: $\bar{\omega}^2 = \omega^2/\alpha$, $\bar{x} = x/A_{1/3}$, $\bar{c} = c/\sqrt{\alpha}$, $\bar{\beta} = \beta A_{1/3}^2/\alpha$, $\bar{F} = F/\alpha A_{1/3}$, $\bar{G} = G/\alpha A_{1/3}$, $\bar{H} = H/\alpha A_{1/3}$, $\bar{A}_1 = A_1/A_{1/3}$, $\bar{B}_1 = B_1/A_{1/3}$. In terms of these quantities (8.1) and (8.2) become

$$(8.3) \qquad \bar{\omega}^2 \bar{x}'' + \bar{c}\bar{\omega}\bar{x}' + (\bar{x} + \bar{\beta}\bar{x}^3) = \bar{H} \cos \theta - \bar{G} \sin \theta,$$

$$(8.4) \qquad \bar{x} = \cos \frac{\theta}{3} + \bar{A}_1 \cos \theta + \bar{B}_1 \sin \theta + \cdots.$$

We now introduce (8.4) in (8.3), make use of the trigonometric identities (7.3) of the preceding section, and obtain (after dropping temporarily the bars over all quantities) the following equations

$$\left(1 - \frac{\omega^2}{9}\right) + \tfrac{3}{4}\beta[1 + A_1 + 2A_1^2 + 2B_1^2] = 0$$

$$(8.5) \qquad\qquad\qquad -\tfrac{1}{3}c\omega + \tfrac{3}{4}\beta B_1 = 0$$

$$(1 - \omega^2)A_1 + \tfrac{1}{4}\beta[1 + 6A_1 + 3A_1^3 + 3A_1 B_1^2] + c\omega B_1 = H$$

$$(1 - \omega^2)B_1 + \tfrac{3}{4}\beta B_1[2 + B_1^2 + A_1^2] - c\omega A_1 = G.$$

In deriving these equations only the two lowest harmonics have been

* Actually, it might have been wise to introduce these quantities much earlier.

taken into account. The first of the equations can be written in the form

(8.6) $\omega^2 - 1 = 8 + \frac{27}{4}\beta[1 + A_1 + 2A_1^2 + 2B_1^2] + \cdots$.

By using the second equation of (8.5) together with (8.6) we can eliminate ω^2 and $c\omega$ from the third and fourth equations of (8.5) to obtain

(8.7) $8A_1 + \frac{1}{4}\beta[-1 + 21A_1 + 27A_1^2 - 9B_1^2 + 51A_1^3 + 51A_1B_1^2] + \cdots = -H$

(8.8) $8B_1 + \frac{3}{4}\beta[7 + 12A_1 + 17A_1^2 + 17B_1^2] + \cdots = -G.$

The above relations coincide with those of the preceding section if we set $c = B_1 = 0$.

We can now introduce A_1 and B_1 as given by (8.7) and (8.8) in (8.6) to obtain the response relation for ω analogous to (7.8) of the preceding section; in terms of the original variables this takes the form

(8.9) $\omega^2 = 9\alpha + \frac{27}{4}\beta\left[A_{1/3}^2 - \frac{FA_{1/3}}{8\alpha} + \frac{H^2}{32\alpha^2}\right].$

We shall not discuss the response curves here, but turn rather to an important observation with regard to the effect of damping. If we square and add relations (8.7) and (8.8) we obtain $64A_1^2 + 64B_1^2 + \cdots = H^2 + G^2 = F^2$, in which the dots refer to terms of order β or higher. It follows therefore that $B_1^2 < F^2/64$ and hence from the second equation of (8.5) that $c\omega < \frac{9}{32}|\beta F|$. In terms of the original variables the last inequality, in view of (8.9), takes the form

(8.10) $c < \frac{3}{32}\frac{|\beta A_{1/3}F|}{\alpha\sqrt{\alpha}}.$

This inequality shows that the subharmonic of order $\frac{1}{3}$ cannot occur unless the damping coefficient is a small quantity of order β, which is perhaps not surprising in view of the remarks made at the beginning of the preceding section. The damping coefficient must also be smaller for smaller "amplitudes" $A_{1/3}$ of the response as well as for smaller amplitudes F of the excitation. It seems likely (cf. Levenson [23]) that the damping coefficient would have to be taken small of still higher order in β in order to obtain the subharmonics of order higher than $\frac{1}{3}$ in the presence of damping. This is an interesting point which deserves further theoretical and experimental investigation.

9. The method of Rauscher

The methods considered so far have consisted in developments in the neighborhood of the free or forced linear oscillation, i.e., in developments which begin, say, at point 0 in Figure 9.1. It is rather obvious that more rapid convergence could be expected if one were to begin the approximations at point 1, i.e., if one were to begin with the free *nonlinear* vibration. The method of Rauscher [35] is an iteration method based on this idea. In applying this method it is just as essential as before to prescribe the amplitude and leave the frequency to be determined. Hence there is some advantage in

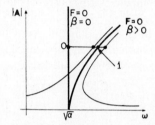

Fig. 9.1. Diagram indicating the character of Rauscher's approximation method.

introducing the variable $\theta = \omega t$ in place of t, as was done in the preceding sections. We therefore take as differential equation

$$(9.1) \qquad \omega^2 \frac{d^2 x}{d\theta^2} + f(x) = F \cos \theta.$$

For the sake of simplicity we assume the restoring force $f(x)$ to be symmetrical, i.e. that $f(-x) = -f(x)$, although Rauscher's method can be carried out without difficulty in other cases. We require that $x(0) = A$, $x'(0) = 0$, which we may do since $x(\theta)$ is assumed to be periodic. The quantity ω is to be fixed by the requirement that $x(\theta)$ should have the period 2π.

As stated above, the Rauscher method is an iteration method which begins with the free nonlinear oscillation as first approximation, that is, with the solution of

$$(9.2) \qquad \omega_0^2 \frac{d^2 x}{d\theta^2} + f(x) = 0,$$

such that $x = x_0(\theta)$ has the period 2π and $x_0(0) = A$, $x_0'(0) = 0$. As we have seen in Chapter II, this problem can be solved by explicit integration: we set $F(x) = \int_0^x f(x)\,dx$ and obtain θ as a function of x by the formula (cf. equation (3.4) of Chapter II):

$$(9.3) \qquad \theta = \theta_0(x) = \omega_0 \int_A^x \frac{dx}{\sqrt{2[F(A) - F(x)]}}.$$

Since we have assumed that $f(-x) = -f(x)$, we obtain for ω_0 the relation

$$(9.4) \qquad \omega_0^{-1} = \frac{2}{\pi} \int_0^A \frac{dx}{\sqrt{2[F(A) - F(x)]}}$$

since $|\theta|$ varies from 0 to $\pi/2$ when x varies from 0 to A.

Once ω_0 and $\theta_0(x)$ have been found, we proceed to determine the next approximation from the differential equation

$$(9.5) \qquad \omega_1^2 \frac{d^2 x}{d\theta^2} + f(x) - F \cos[\theta_0(x)] = 0,$$

in which $\theta_0(x)$ is taken from (9.3). In other words, we go back to (9.1) and replace θ in the right-hand side by its first approximation $\theta_0(x)$. This process is quite feasible since (9.5) is again a differential equation of the same type as (9.2) and can, like it, be solved by explicit integration to obtain the next approximation ω_1 for the frequency and $\theta_1(x)$ for the "time" as a function of x. In complicated cases the integrations are best carried out graphically, as Rauscher illustrates in his paper.

One sees that the general scheme for this iteration method is as follows: The nth approximation $x_n(\theta)$ is obtained as the solution of

$$(9.6) \qquad \omega_n^2 \frac{d^2 x}{d\theta^2} + f(x) - F \cos[\theta_{n-1}(x)] = 0$$

for which $x_n(\theta) = x_n(\theta + 2\pi)$ and $x_n(0) = A$, $x_n'(0) = 0$. The function $\theta_{n-1}(x)$ is the inverse of the solution $x_{n-1}(\theta)$ obtained in the preceding step. (Actually, of course, it is $\theta_{n-1}(x)$ which is obtained directly, and this is what makes this ingenious method of Rauscher practicable.) The quantity ω_n is the nth approximation to the frequency of the forced oscillation of amplitude A. The accuracy of this iteration method depends mainly, as one sees, on having F sufficiently small.

The method seems to be a very accurate one, as Rauscher indicates in his paper; in fact, it is usually sufficient to stop with x_1 . In the paper referred to, it is also indicated how the method can be modified if a damping term $c\dot{x}$ occurs in the differential equation. It should be mentioned that the method of Rauscher involves an assumption that was not mentioned explicitly above, i.e. that the curve for $x(\theta)$ should be monotone over a full period: otherwise, the inverse function $\theta(x)$ would not exist.

10. Combination tones

Up to now we have always considered the excitation to consist of a single harmonic. It is, however, a matter of considerable interest from the practical, and perhaps even more from the mathematical, point of view to investigate what occurs when the excitation is a sum of several harmonics. We consider therefore the differential equation

$$(10.1) \qquad \ddot{x} + \alpha x - \beta x^3 = H_1 \cos \omega_1 t + H_2 \cos \omega_2 t.$$

To this differential equation we might apply any one of several of the approximation schemes used earlier in this chapter. For example, the iteration method of Duffing explained in Section 2 might be used, starting with

$$(10.2) \qquad x_0 = A \cos \omega_1 t + B \cos \omega_2 t$$

as first approximation. In this way Duffing [9], p. 111, obtained the following result for the next approximation $x_1(t)$ (in slightly different notation and with sine rather than cosine terms):

$$
\begin{aligned}
x_1 = {}& A \cos \omega_1 t + B \cos \omega_2 t + \frac{\beta A^3}{36} \cos 3\omega_1 t + \frac{\beta A^3}{36} \cos 3\omega_2 t \\
(10.3) \qquad & + \tfrac{3}{4}\beta A^2 B \left\{ \frac{\cos (\omega_2 + 2\omega_1)t}{(\omega_2 + 2\omega_1)^2} + \frac{\cos (\omega_2 - 2\omega_1)t}{(\omega_2 - 2\omega_1)^2} \right\} \\
& + \tfrac{3}{4}\beta A B^2 \left\{ \frac{\cos (\omega_1 + 2\omega_2)t}{(\omega_1 + 2\omega_2)^2} + \frac{\cos (\omega_1 - 2\omega_2)t}{(\omega_1 - 2\omega_2)^2} \right\},
\end{aligned}
$$

in which A and B must satisfy the relations

$$(10.4) \qquad \begin{cases} (\alpha - \omega_1^2)A - \tfrac{3}{4}\beta A(A^2 + 2B^2) = H_1, \\ (\alpha - \omega_2^2)B - \tfrac{3}{4}\beta B(B^2 + 2A^2) = H_2. \end{cases}$$

The relations (10.4) are, of course, the response relations for the amplitudes A, B as functions of the frequencies ω_1, ω_2.

One of the essential points of interest here is the occurrence of terms of first order in β with frequencies $\omega_2 \pm 2\omega_1$ and $\omega_1 \pm 2\omega_2$, the so-called combination tones, in addition to those of frequency $3\omega_1$ and $3\omega_2$. This is in sharp contrast with the case of linear oscillations in which the forced oscillation would be simply a superposition of two harmonics of frequencies ω_1 and ω_2. If we had taken a nonlinear term of the form $-\gamma x^2$ in the right-hand side of (10.1) instead of the term $-\beta x^3$ a similar result (cf. Duffing [9], p. 108, and Appendix II) would have been obtained except that the combination tones of lowest order would now consist of the "difference tone" of frequency $\omega_2 - \omega_1$ and the "summation tone" of frequency $\omega_2 + \omega_1$. Helmholtz invoked a supposed nonlinearity in the mechanical vibrating system of the ear as a means of explaining, along lines similar to the above discussion, the fact that tones of frequencies $\omega_2 - \omega_1$ and $\omega_2 + \omega_1$ are often heard when two notes of frequencies ω_1 and ω_2 are sounded, especially if they are loud notes. An excellent summary and critical discussion of such acoustical phenomena can be found in Rayleigh [36], p. 456.

We turn now to a brief discussion of solutions of (10.1) from the mathematical point of view. If ω_1/ω_2 is rational, then the excitation $H_1 \cos \omega_1 t + H_2 \cos \omega_2 t$ is periodic and the method of Appendix I could be applied to furnish the existence of periodic solutions of (10.1) of various kinds which could be represented by convergent perturbation series. If, however, ω_1/ω_2 is irrational and the excitation is thus an almost-periodic function of the time, the situation is quite different. The strange and noteworthy fact is that the determination of solutions of (10.1) by the kind of approximation methods used hitherto—iteration or perturbation methods of one sort or another—seems certain to lead to series or sequences which *diverge if the ratio ω_1/ω_2 of the frequencies ω_1 and ω_2 is not a rational number.* It is possible to understand why this should be so by considering what would happen in the present case of equation (10.1) if, for example, the iteration scheme of Duffing were to be continued in order to obtain approximations of higher order. Since the basic process consists essentially in substituting each approximation x_n in the right-hand side of $\ddot{x} = \beta x_n^3 - \alpha x_n + H_1 \cos \omega_1 t + H_2 \cos \omega_2 t$ and integrating twice to obtain x_{n+1}, it follows (as we observe already in the case of $x_1(t)$

given by (10.3)) that the resulting approximations will contain more and more terms whose denominators contain higher and higher powers of $(\pm n\omega_1 \pm m\omega_2)$ with n and m integers. By virtue of a famous theorem of Kronecker, however, it is known that expressions of the form $(\pm n\omega_1 \pm m\omega_2)$ with n and m integral will be arbitrarily near to zero for infinitely many different n and m if ω_1/ω_2 is irrational. Consequently it would seem quite hopeless to attempt to prove the convergence of any such iteration process as that sketched out here. This difficulty is the famous "difficulty of the small divisors" which was first pointed out by Poincaré in discussing perturbation methods for dealing with problems in celestial mechanics.

The difficulty with the small divisors can, however, be circumvented in some cases at least if there is viscous damping in the system. In Appendix II we shall show, following Friedrichs [40], that the differential equation $\ddot{x} + c\dot{x} + x - \beta x^2 = h(t)$, with $c > 0$ and $h(t)$ an almost periodic function, possesses almost periodic solutions in a neighborhood of $x \equiv 0$ which can be obtained by a convergent iteration process for sufficiently small values of β. The essence of the matter is that the occurrence of viscous damping has the effect of causing the denominators in the successive approximations to be bounded away from zero. The fact that actual systems are not observed to blow up when subjected to the influence of harmonics with incommensurable frequencies—in spite of the fact that series for describing the phenomena seem certain to diverge—could, however, hardly be ascribed in general to the effect of damping.

11. Stability questions

The jump phenomena discussed in Section 3 make it seem plausible that the solutions corresponding to values of $|A|$ and ω for those portions of the harmonic response curves which are passed over in jumping from one point to another are unstable solutions in some sense or other. In other words, it seems plausible that the regions bounded by the dotted curves in Figure 3.1 are regions of instability for the equation $\ddot{x} + c\dot{x} + \alpha x + \beta x^3 = F \cos \omega t$.

In any consideration of stability of a given system one fundamental difficulty is that of defining the notion of stability in a logical and reasonable manner without destroying the chances of applying the definition in a practical way. Into this question we do not enter here. We take simply the following often used definition for stability (infinitesimal stability): Let $x(t)$ and $x(t) + \delta x(t)$ be two solutions of the differential equation of motion for which the initial conditions at $t = 0$, say, differ slightly. If we insert $x + \delta x$ in the differential equation and neglect powers of δx above the first, we obtain a linear "variational" differential equation for δx. *If all solutions δx of this equation are bounded, then $x(t)$ is said to be stable, otherwise unstable.*

When this definition is applied to the Duffing equation (1.3) we find for $\delta x(t)$ the linear homogeneous differential equation

$$(11.1) \qquad \delta \ddot{x} + (\alpha + 3\beta x^2)\delta x = 0,$$

in which $x(t)$ is the solution of (1.3) whose stability is to be investigated. In our case $x(t)$ is a periodic function, so that (11.1) is what is called a Hill's equation. To each $x(t)$, or, to each point in the $|A|$, ω-plane, corresponds a particular equation (11.1). In any specific case means are available to decide whether all solutions δx are bounded or not. However, it seems not to have been proved rigorously that all solutions corresponding to points in the region bounded by the dotted curves of Figure 3.1 are unstable. A strong indication that this is true is furnished by the fact that it can be shown that (11.1) always possesses a *periodic* solution $\delta x(t)$ for any $x(t)$ which corresponds to a boundary point of the regions in question. It follows then from certain theorems of O. Haupt that such points in the $|A|$, ω-plane correspond to boundary points between regions of stable and unstable solutions.*

If one were to be contented with the degree of accuracy implied in the first approximation $x = A \cos \omega t$ to the solution of Duffing's equation, equation (11.1) would take the form

$$(11.2) \qquad \delta \ddot{x} + (a + b \cos 2\omega t)\delta x = 0,$$

with a and b defined as follows:

$$(11.3) \qquad \begin{aligned} a &= \alpha + \tfrac{3}{2}\beta A^2, \\ b &= \tfrac{3}{2}\beta A^2. \end{aligned}$$

* F. John has shown [18], [19] that these statements are correct as regards the *out-of-phase branch* of the response curves for harmonic solutions which are not necessarily in a neighborhood of the linear solution.

For later purposes it is useful to introduce the following new quantities

(11.4)
$$\begin{cases} z = 2\omega t \\ \delta = a/4\omega^2 \\ \epsilon = b/4\omega^2, \end{cases}$$

in terms of which (11.2) takes the form

(11.5)
$$\frac{d^2 \delta x}{dz^2} + (\delta + \epsilon \cos z)\delta x = 0.$$

Equation (11.5), which is of course a special case of Hill's equation, is called the Mathieu equation. The stability theory of this equation has been worked out in detail for all values of ϵ and δ. In Chapter VI we shall develop the theory of Hill's equation and the Mathieu equation in detail and apply it to the discussion of the stability of the harmonic solutions of Duffing's equation. In particular, it will be shown there that the statement made above regarding the unstable regions of the $|A|$, ω-plane holds good, at least within first order terms in β.

12. Résumé

The frequently cited book of Duffing contains in one place a brief confrontation, in the form of a table, of the characteristics and properties of linear as contrasted with nonlinear systems. We close this chapter with a similar and somewhat enlarged table as a means of summarizing the salient facts discussed up to this point.

Free Oscillations: $\ddot{x} + c\dot{x} + x + \beta x^3 = 0, c > 0$

Linear ($\beta = 0$)	Nonlinear ($\beta \neq 0$)
(1) $c \neq 0$: $x \equiv 0$ only periodic motion.	(1) $c \neq 0$: $x \equiv 0$ only periodic motion.
(2) $c = 0$: Simple harmonic motion with arbitrary amplitude but fixed period $T = 2\pi$.	(2) $c = 0$: Motions periodic but not simple harmonic. Period T is a unique function of amplitude, and $T \neq 2\pi$ except for $\beta = 0$.

Forced Oscillations: $\ddot{x} + c\dot{x} + x + \beta x^3 = H(t)$

Linear ($\beta = 0$)	Nonlinear ($\beta \neq 0$)
(1) $c = 0$, $H(t) = F \cos \omega t$ $\omega \neq 1$: All oscillations a superposition of free oscillation of period 2π and arbitrary amplitude, and a forced oscillation $A \cos \omega t$ with A fixed. All oscillations stable. $\omega = 1$: Resonance case. No periodic oscillation. Forced oscillation has "amplitude" increasing linearly in t.	(1) $c = 0$, $H(t) = F \cos \omega t$ There exist oscillations of frequency ω, with amplitudes a function of ω. For certain ω several such oscillations may occur. For $\omega = 1$ (for which resonance occurs in linear case) stable oscillations of frequency $\omega = 1$ with bounded amplitudes occur. General solution not known, but there exist many different types of periodic motions (subharmonics, etc.) as well as nonperiodic motions. Some periodic oscillations stable, others unstable. Behavior of periodic oscillations depends considerably on sign of β.
(2) $c \neq 0$, $H(t) = F \cos \omega t$ All oscillations a superposition of a free oscillation, which is damped out exponentially, and a forced oscillation $A \cos \omega t$ with A fixed. All motions stable.	(2) $c \neq 0$, $H(t) = F \cos \omega t$ Essentially the same remarks as for $c = 0$.
(3) $c \neq 0$, $H(t) = F_1 \cos \omega_1 t + F_2 \cos \omega_2 t$ ω_1, $\omega_2 \neq 1$ All oscillations a superposition of a damped free oscillation and a forced oscillation which is in turn a superposition of the forced oscillation due to each of the two separate terms in $H(t)$ individually.	(3) $H(t) = F_1 \cos \omega_1 t + F_2 \cos \omega_2$ $c = 0$: General motion unknown. Formal approximation schemes lead to combination tones. Mathematical "difficulty of small divisors" occurs. $c \neq 0$: Existence of certain oscillations containing combination tones can be proved.

CHAPTER V

Self-sustained Oscillations

In the preceding chapter we treated problems in which the nonlinear element in the system was the restoring force. In the present chapter we consider problems in which the nonlinear element concerns the force depending on the velocity, while the restoring force is assumed to be linear. These nonlinear "damping" forces will have, however, a special and very important property in all cases we consider: the damping force will be such as to tend to *increase the amplitude for small velocities but to decrease it for large velocities.* It follows that the state of rest is not a stable state in such cases and that an oscillation will be built up from rest even in the absence of external forces; this accounts for the description of these oscillations as self-excited or self-sustained oscillations. The present chapter is divided into two parts A and B; in the first part the free oscillations and in the second the forced oscillations are treated.

A. Free Oscillations

1. *An electrical problem leading to free self-sustained oscillations*

Probably the most important physical systems in practice which lead to oscillations of the type to be considered in this chapter are electrical systems involving vacuum tubes. An electrical circuit containing such an element is shown in Figure 1.1. The triode vacuum tube indicated in the figure contains three main elements: the plate (or anode) P, the grid G, and the cathode (or filament) F. The cathode is heated by a small battery, as indicated, in order that it will emit electrons. If, then, the anode P is charged positively an electric field will be created in the tube and an electron current will flow from cathode to plate. Between the plate and the cathode an additional element, the grid, is interposed; the grid, as the name indicates, consists of a wire screen of coarse mesh. The grid was introduced into vacuum tubes for the purpose of controlling the flow of electrons. This is accomplished simply by changing the potential of the grid, thus modifying the electric field in the tube and hence also the flow of electrons. The control can be achieved without an appreciable flow of current in the grid circuit, if desired. It is this central control feature (which requires only slight expenditure of energy) that has made the vacuum tube such a useful device for a great variety of purposes.

The circuit shown in Figure 1.1 was chosen because it is important in practice and leads easily to the differential equation that we wish to

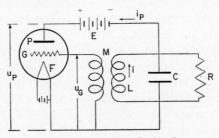

FIG. 1.1. A feed-back circuit.

study. The distinguishing feature of the physical phenomenon is also well illustrated: by means of a nonlinear element (in this case the triode tube) a source which normally provides a constant flow of energy (the battery in the plate circuit) is forced to produce oscillations. It should be pointed out specifically that such occurrences require, in general, the presence of nonlinear elements: in a linear system a periodic source of energy is required in order to maintain a periodic motion (cf. the paper of le Corbeiller [7] for interesting comments on this point).

In the plate circuit there is, in addition to the battery, an "oscillator" consisting of a coil of inductance L, a resistance R, and a condenser of capacity C all in parallel. The grid potential is provided by a mutual inductance M, as indicated. Such a circuit is sometimes called a feed-back circuit, in contrast with other circuits in which the grid control is provided by a circuit that is independent of the plate circuit.

We proceed to derive the differential equation for the current i flowing through the inductance coil of the oscillator. As indicated earlier, we assume that the current in the grid circuit may be neglected. The drop of potential through the inductance coil is, therefore, $L di/dt$ and this is also the potential drop through the resistance and the condenser. If we denote by i_R the current through the resistance, by i_C the current through the condenser, and by q_C the charge on the condenser, we have the following relations:

(1.1)
$$L\frac{di}{dt} = Ri_R$$

(1 2)
$$L\frac{di}{dt} = \frac{1}{C}q_C \quad \text{or} \quad L\frac{d^2 i}{dt^2} = \frac{1}{C}\frac{dq_C}{dt} = \frac{1}{C}i_C$$

(1.3)
$$i_P = i + i_R + i_C .$$

If the quantities i_R and i_C in (1.3) are replaced by their values from (1.1) and (1.2) the result is

(1.4)
$$CL\frac{d^2 i}{dt^2} + \frac{L}{R}\frac{di}{dt} + i = i_P .$$

Since no current was assumed to flow in the grid circuit it follows that the grid potential u_G is given by

(1.5)
$$u_G = M\frac{di}{dt}.$$

Also we have

(1.6)
$$u_P = E - L\frac{di}{dt},$$

u_P being the plate potential.

So far we have made practically no use of the properties of the tube itself. The decisive point is that the plate current i_P depends, with good accuracy, upon a linear combination u of the grid potential u_G and the plate potential u_P :

(1.7)
$$i_P = \varphi(u)$$

with

(1.8)
$$u = u_G + Du_P , \qquad D > 0,$$

D being a certain constant, the reciprocal of which is called the "amplification factor." Equation (1.7) might be given a theoretical justification, but we regard it as an empirical relation that is justified by the results of experiments. The function $\varphi(u)$ is sometimes called the characteristic of the tube.

If $\varphi(u)$ were a linear function of u we see that equations (1.4) to (1.8) inclusive would lead to a linear differential equation with constant coefficients for i, so that an oscillation of some kind could be expected to occur. Much of the engineering literature on this subject is based on the assumption that $\varphi(u)$ can be taken as a linear function without too much error.

The function $\varphi(u)$ is not linear, but has rather the character indicated in Figure 1.2. That the characteristic is a curve of this kind was to be expected, since the current i_P is limited finally by the

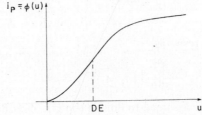

FIG. 1.2. Plate current as a function of grid and plate potentials.

rate of production of electrons at the cathode; the curve should then show evidence of such a saturation effect.

From (1.5), (1.6), and (1.8) we have

$$(1.9) \qquad u = DE + (M - DL)\frac{di}{dt},$$

and, since $i_P = \varphi(u)$ is nonlinear in u we see that (1.4) is a nonlinear differential equation in which the nonlinearity involves the first derivative.

For the discussion of the differential equation it is convenient to introduce a new dependent variable x defined by

$$(1.10) \qquad x = i - \varphi(DE).$$

In the new variable, equation (1.4) becomes

$$(1.11) \quad CL\ddot{x} + \frac{L}{R}\dot{x} + x + \varphi(DE) = \varphi[DE + (M - DL)\dot{x}],$$

in which $\dot{x} = dx/dt = di/dt$. Through introduction of the function $f(\dot{x})$ defined by

(1.12) $$f(\dot{x}) = \varphi(DE) - \varphi[DE + (M - DL)\dot{x}],$$

we may write (1.11) in the form

(1.13) $$CL\ddot{x} + \frac{L}{R}\dot{x} + f(\dot{x}) + x = 0.$$

We suppose now that the plate battery potential E is adjusted so that the quantity $u = DE$ is near the inflection point of the characteristic $\varphi(u)$ (cf. Figure 1.2). We assume, in addition, that $M - DL > 0$; this condition is very essential—without it no self-

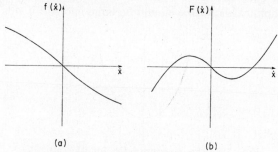

(a)

(b)

FIG. 1.3. The characteristic for the circuit.

excited oscillation would be possible, as we shall see. This condition is certainly satisfied if D is small enough, i.e., if the amplification factor $1/D$ is large enough. The curve for $f(\dot{x})$ as a function of $\dot{x}$ then has the general appearance indicated in Figure 1.3a. In Figure 1.3b we indicate the curve for the function $F(\dot{x})$—called the *characteristic*—defined by

(1.14) $$F(\dot{x}) = \frac{L}{R}\dot{x} + f(\dot{x}),$$

assuming that the slope of $F(\dot{x})$ for $\dot{x} = 0$ is negative, i.e., that

(1.15) $$\frac{L}{R} + f'(0) < 0,$$

or

(1.15)′ $$\frac{L}{R} < (M - DL)\varphi'(DE),$$

which can be secured, in view of $L/R > 0$, only if $f'(0) < 0$, that is, if $M - DL > 0$. We may now write (1.13) in the form

$$(1.16) \qquad CL\ddot{x} + F(\dot{x}) + x = 0,$$

with $\dot{x}F(\dot{x}) < 0$ for $|\dot{x}|$ small enough, but $\dot{x}F(\dot{x}) > 0$ for large $|\dot{x}|$. This means that the "damping" is negative for small $\dot{x}$; the system absorbs energy, and one could expect the amplitude of x to increase. However, for $\dot{x}$ large the system dissipates energy and hence one could expect the amplitude of x to be limited from above finally. In other words, *one could expect under these circumstances that a steady vibration of a certain amplitude would occur after some transients die out.* We note that it is quite essential for such a behavior of the system that the slope of $F(\dot{x})$ should be negative for small $\dot{x}$, or, in other words that the amplification factor should be large enough; if this were not

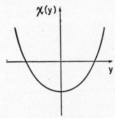

Fig. 1.4. The characteristic for the circuit.

the case no oscillation (except the state $x \equiv 0$) could occur, as we have seen earlier in Chapter III, Section 2.

A differential equation of the form (1.16) was first studied by Rayleigh [36] in connection with acoustical problems. The treatment of the important problems centering about circuits with vacuum tubes in terms of differential equations with nonlinear damping is of much more recent date; this pioneering work was done by van der Pol [32].

Many writers, including van der Pol, prefer to work with a differential equation different from (1.16). Instead of x, they take as variable its derivative $y = \dot{x}$. If (1.16) is differentiated with respect to t the result is

$$CL\ddot{x} + F'(\dot{x}) \cdot \ddot{x} + \dot{x} = 0,$$

or

$$(1.17) \qquad CL\ddot{y} + \chi(y) \cdot \dot{y} + y = 0$$

upon setting $y = \dot{x}$ and $\chi(y) = F'(y)$.   If $F(\dot{x})$ is of the type indicated
in Figure 1.3b, we see that $\chi(y)$ appears as in Figure 1.4.   If $\chi(y)$
is of this general character we may present once more an argument
which indicates that a self-sustained oscillation will occur: For
small $|\,y\,|$, the coefficient $\chi(y)$ of the damping term is negative and
hence $|\,y\,|$ will tend to increase.   For large enough $|\,y\,|$, the coefficient
of y becomes and remains positive so that $|\,y\,|$ will tend to decrease.
The interplay of these two effects of opposite tendency might be
expected to result in a steady oscillation of a certain amplitude.

2. Self-sustained oscillations in mechanical systems

There are a number of well known cases of mechanical systems
which execute oscillations of the type under consideration here.   In

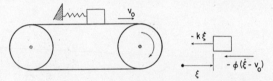

FIG. 2.1. Mechanical system capable of executing self-sustained oscillations.

one class of such problems the oscillations result from the action of
solid friction.   Consider, for example, the mechanical system shown
in Figure 2.1.   A block rests on a rough belt which moves with a
constant speed v_0 , and is attached to a rigid support through a linear
spring.   If the speed v_0 of the belt is properly chosen, it is a well
known fact of experience that the block will not remain at rest, but
will instead execute somewhat jerky oscillations.   This behavior
results from the fact that the solid friction force (or Coulomb damping,
as it is also called) between the block and the belt is not constant,
but varies with the velocity of slipping $\dot{s}$ of the block relative to the
belt.   In Figure 2.2 we indicate schematically the manner in which
the friction force $-\varphi(\dot{s})$ varies with $\dot{s}$.   When the block is at rest
relative to the belt ($\dot{s} = 0$), the friction force increases numerically
(when an external force tending to move it is applied to the block)
until a certain critical value is reached, after which it decreases for a
time only to become larger again when $\dot{s}$ becomes large.   It is clear
that the friction force and $\dot{s}$ have opposite signs.

The position of the block is assumed to be fixed by its distance ξ from the point at which the spring is neither stretched nor compressed. The velocity of slip $\dot{s}$ is then given by

$$(2.1) \qquad\qquad \dot{s} = \dot{\xi} - v_0 ,$$

and the equation of motion of the block is (cf. Figure 2.1):

$$(2.2) \qquad\qquad m\ddot{\xi} + \varphi(\dot{\xi} - v_0) + k\xi = 0.$$

It is convenient to introduce a new variable x, replacing ξ, by the relation

$$(2.3) \qquad\qquad x = \xi + \frac{1}{k}\varphi(-v_0),$$

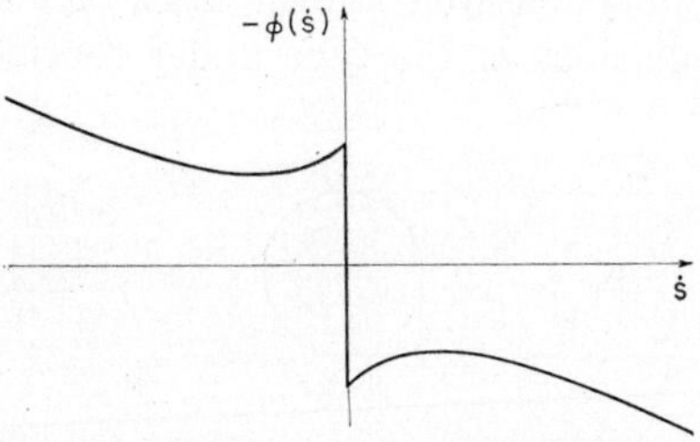

FIG. 2.2. Damping force as a function of velocity.

which means that the position of the block is now measured from its equilibrium position under the combined action of the spring force and the friction force, since $\varphi(-v_0) + k\xi = 0$ characterizes this position, in view of (2.2). The differential equation for x is then

$$(2.4) \qquad m\ddot{x} + [\varphi(\dot{x} - v_0) - \varphi(-v_0)] + kx = 0,$$

or

$$(2.5) \qquad\qquad m\ddot{x} + F(\dot{x}) + kx = 0$$

with $F(\dot{x})$ defined by

$$(2.6) \qquad\qquad F(\dot{x}) = \varphi(\dot{x} - v_0) - \varphi(-v_0).$$

The function $F(\dot{x})$ will appear as in Figure 2.3 *if v_0 is not taken too large*, since the function φ behaves as indicated in Figure 2.2; in particular, it is important that the slope of this curve is negative at the origin. One sees that the latter requirement will be fulfilled only if v_0 is such that the friction force between block and belt would decrease numerically if v_0 were increased. The importance of this for the physical problem is of course that self-excited oscillations can occur only when the damping force behaves in this way for small values of $\dot{x}$, as we might infer from the discussion in the preceding section.

The specific problem in mechanics chosen here is only one possible case which leads to a self-excited oscillation. The production of oscillations in a violin string which result from drawing a bow in one direction across the string was first explained by Rayleigh in much the same manner as above (i.e., by taking account of the variability of the solid friction between the bow and the string). Self-excited oscillations due to the same cause have been observed in a pendulum

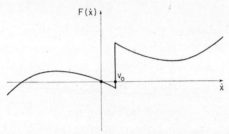

Fig. 2.3. The characteristic for the mechanical system.

swinging on a rotating shaft. Whirling of a shaft caused by solid friction in a loose bearing has also been explained in this way. Still another example of a self-excited oscillation caused by solid friction is the often observed "chattering" of the brake shoes against the wheels of a railroad car when the brakes are applied.

Self-excited oscillations sometimes occur in electrical transmission line wires due to the action of the wind. The failure of the Tacoma bridge is ascribed to a very heavy oscillation of the same type. However, the description of the type of action which underlies phenomena of this sort is somewhat complicated, because of the necessity to consider the flow pattern of the air behind the structure, and will not be undertaken here. That these latter cases (and others, including flutter of aeroplane wings) really belong to the class of self-excited oscillations is fairly clear on general grounds since the vibration results through partial conversion of energy from a steady flow (in these cases the wind) into oscillations.

3. A special case of the van der Pol equation

We consider once more the equation (1.16):

$$(3.1) \qquad CL\ddot{x} + F(\dot{x}) + x = 0.$$

If it is considered that the vacuum tube operates near the point of inflection of the *characteristic* $F(\dot{x})$, the function $F(\dot{x})$ can be approximated with good accuracy by the expression

$$(3.2) \qquad F(\dot{x}) = -\alpha\dot{x} + \frac{\beta}{3}\dot{x}^3, \qquad \alpha, \beta > 0,$$

for $\dot{x}$ not too large. The differential equation then becomes

$$(3.3) \qquad CL\ddot{x} + \left(-\alpha\dot{x} + \frac{\beta}{3}\dot{x}^3 \right) + x = 0.$$

Van der Pol observed that (3.3) could be given a much simpler form involving only one parameter by introducing the new quantities

$$(3.4) \qquad t_1 = \omega t, \quad \omega^2 = 1/CL \quad x_1 = (\omega\sqrt{\beta/\alpha})x,$$

and a new parameter ϵ through the relation

$$(3.5) \qquad \epsilon = \alpha/\sqrt{CL} = \alpha\omega,$$

in terms of which the differential equation (3.3) is transformed into

$$(3.6) \qquad \frac{d^2x_1}{dt_1^2} - \epsilon\left[\frac{dx_1}{dt_1} - \frac{1}{3}\left(\frac{dx_1}{dt_1} \right)^3 \right] + x_1 = 0.$$

As one sees, the parameter ϵ occurs as the coefficient of the damping term.

A considerable portion of the remainder of this chapter will be concerned with the discussion of equation (3.6). Hence it is convenient to drop the subscripts and take as the fundamental differential equation for our later discussion the equation

$$(3.7) \qquad \ddot{x} + \epsilon F(\dot{x}) + x = 0, \qquad \epsilon > 0$$

with $F(\dot{x})$ defined most frequently by

$$(3.8) \qquad F(\dot{x}) = -\dot{x} + \tfrac{1}{3}\dot{x}^3.$$

4. The basic character of self-excited oscillations

The differential equation (3.7) is one governing what have been called earlier "free" oscillations, since the time t does not occur explicitly. Hence we may apply to it the methods discussed earlier in Chapter III. Upon introduction of the "velocity" $v = \dot{x}$ as new variable (3.7) reduces to the first order differential equation

$$(4.1) \qquad \frac{dv}{dx} = \frac{\epsilon(v - v^3/3) - x}{v} = \frac{-\epsilon F(v) - x}{v}, \qquad \epsilon > 0.$$

The field direction defined by (4.1) at any point in the x, v-plane can be readily determined, as we have already seen in Chapter III, Section 3, by a graphical construction due to Liénard. For the sake of convenience we repeat the description of this construction here. The method is indicated in Figure 4.1. The characteristic $x = -\epsilon F(v)$ is first plotted. The field direction at $P(x, v)$ is then obtained as

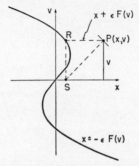

FIG. 4.1. The Liénard construction.

follows. From P a line is drawn parallel to the x-axis until it cuts the characteristic at R. From R a perpendicular is dropped to the x-axis at S; the field direction at P is then orthogonal to the line SP. That the construction is correct follows at once from (4.1) since the slope of SP is obviously given by $v/[x + \epsilon F(v)]$.

In the special case $\epsilon = 0$, the point S is the origin and the integral curves of (4.1) are concentric circles with their common center at the origin; the motion is a simple harmonic motion. The only singularity is a center at the origin. For $\epsilon > 0$ the singularity becomes an unstable spiral point (see Chapter III, Section 6); the integral curves depart from the origin for $t > 0$, corresponding to the fact that the origin is an unstable equilibrium point. Our previous physical arguments indicate, however, that the outward spirals near the origin will not spread out indefinitely, since the damping becomes positive for large values of v. Far away from the origin the integral curves are in fact spirals which turn about the origin in moving toward it as t increases. Two such spirals are shown in Figure 4.2, which was obtained for $\epsilon = 1$ by applying the construction of Liénard stepwise. The figure indicates clearly what happens. *The spirals near the origin as well as those far away from the origin tend to a single closed integral curve of* (4.1), which in its turn corresponds to a periodic solution of (3.7). In other words, *every solution of* (3.7) *tends, as* $t \to +\infty$, *to a periodic solution.* These are the salient facts about self-excited oscillations. Occurrences of this kind were first studied by Poincaré, who gave the name *limit cycle* (or simply *cycle*) to a closed solution curve of the kind we have found in the present case.

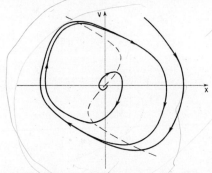

Fig. 4.2. The limit cycle for the van der Pol equation.

We note that the closed cycle contains one singular point of index $+1$ in its interior, i.e. the spiral point at the origin, and that this is in accord with our discussion in Chapter III, Section 7, according to which the sum of the indices of the singularities inside a closed solution curve should be $+1$.

If ϵ is small the closed solution curve which results will be nearly a circle and the corresponding motion will be nearly a simple harmonic motion of a definite amplitude. As ϵ is increased the limit state that is approached for large t deviates more and more from a simple harmonic motion. Figure 4.3 shows the nature of the curves for $v = \dot{x}$ as a function of t for $\epsilon = 0.1$, $\epsilon = 1$, and $\epsilon = 10$ for a solution which begins with small values of x and v for $t = 0$. The distortion from a sine wave increases markedly with increase of ϵ. The very jerky oscillations which occur for values of ϵ larger than about 10, say, are often called *relaxation oscillations*; we shall have more to say about them later. However, it is of interest to observe here that such relaxation oscillations will occur in the circuit of Figure 1.1 if the capacity C of the condenser in the oscillating circuit is made small, as one sees from (3.5).

It is clear that the same general phenomena will occur whenever the characteristic has a positive slope for small x and v and a negative slope for large x and v. In particular, the problem of the motion of a block lying on a moving belt and subjected to solid friction, which

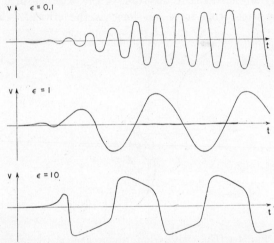

FIG. 4.3. Change in character of the oscillations with increase in nonlinearity.

was formulated in equation (2.5) above, is a case in point. The equation (2.5) can be put in a form suitable for the application of the Liénard construction by introducing a new independent variable t_1 replacing the time t through the relation $t = t_1 \sqrt{m/k}$; the result is

$$(4.2) \qquad \frac{d^2x}{dt_1^2} + \frac{1}{k} F\left(\sqrt{\frac{k}{m}} \frac{dx}{dt_1}\right) + x = 0.$$

Upon introducing $v = dx/dt_1$ equation (4.2) can be reduced, in the usual way, to the following first order equation:

$$(4.3) \qquad \frac{dv}{dx} = \frac{-\frac{1}{k} F\left(\sqrt{\frac{k}{m}} v\right) - x}{v}.$$

In Figure 4.4 we show the result of applying the Liénard construction to (4.3) when the characteristic F has the form indicated in Figure 2.3. It is amusing to observe that once any integral curve of (4.3) touches the straight line segment P_1P_2 it follows it moving from left to right until P_1 is reached, since on this segment $v \neq 0$ and thus $v = \text{constant}$ is a solution of (4.3). (One can also see this in terms of the Liénard construction by thinking of the segment P_1P_2 as the limit of a segment rotated slightly out of the horizontal position.) From the mechanical point of view, what this means is that the solid friction force on the block simply adjusts itself to the value of the external force applied—in this case the spring force—as long as the critical value of the

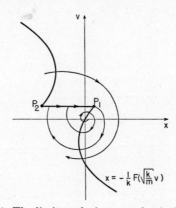

Fig. 4.4. The limit cycle for a mechanical system.

friction force is not exceeded; hence the system moves with constant velocity since the resultant force is zero. To find the limit cycle toward which all other integral curves of (4.3) tend as $t \to \infty$ it was sufficient in the case shown in Figure 4.4 to construct the integral curve which starts at P_1 and to follow it until it touched the segment P_1P_2 for the first time. However, it need not always happen that the limit cycle contains a portion of the segment P_1P_2 : for this to occur it is necessary that the point P_2 lie far enough to the left, and this in turn requires that the critical value of the friction force be not too small. From (4.2) one sees also that the departure from linearity, or from a simple harmonic oscillation, depends essentially on the stiffness of the spring: if k, the spring constant, is large the oscillations will be nearly simple harmonic, but if k is small the oscillations will be of the relaxation type; in other words, jerky oscillations are likely to occur with a weak spring, as one would intuitively expect. The case shown in Figure 4.4 should doubtlessly be classed as a relaxation oscillation. The suppression of the spring thus has the same effect on the oscillations here as the suppression of the condenser in the electrical problem treated above.

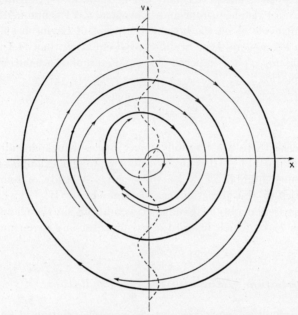

FIG. 4.5. Case in which infinitely many limit cycles occur.

For any given $F(v)$ the graphical solution is rather easily obtained, so that it is not difficult to decide practically whether a closed solution curve exists or not. The Liénard construction can also furnish accurate quantitative results. However, it is not a particularly easy problem to prove rigorously that a closed solution curve exists; and it is still less easy to show that there is a *unique* closed solution curve or cycle. It is in fact quite clear that the solutions of equation (4.1) will have neither of these properties unless $F(v)$ has certain special features. This question has been treated by Liénard [25] and, in a more general way, by Levinson and Smith [24], who considered the differential equation

$$(4.4) \qquad \frac{dv}{dx} = \frac{-f(x, v)v - g(x)}{v}.$$

Levinson and Smith found conditions on $f(x, v)$ and $g(x)$ which insure that a closed solution curve exists, and other more complicated conditions which insure that only one cycle occurs. In Appendix III we given an existence proof and in Appendix VI a uniqueness proof for a limit cycle along the lines of the proof of Levinson and Smith, but for a less general type of differential equation than (4.4).

In Figure 4.5 we indicate the character of the solution curves of the equation

$$(4.5) \qquad \frac{dv}{dx} = \frac{\sin v - x}{v}.$$

In this case there are infinitely many cycles which are alternately stable and unstable. It is also not difficult to give examples in which any specified number of cycles will occur. A study of some of the possibilities in this direction has been made by H. Eckweiler [10], who in particular obtained sufficient conditions on the characteristic in certain cases which insure the existence of any given number of cycles.

5. Perturbation method for the free oscillation

If ϵ is small we can determine the periodic solution of (3.7) using the perturbation method.* Since we do not know the frequency of the periodic solution it is advantageous, as we have seen earlier, to replace the independent variable t by $\theta = \omega t$, ω being the unknown frequency of the periodic solution. The differential equation (3.7), with F defined by (3.8), becomes

$$(5.1) \qquad \omega^2 \frac{d^2x}{d\theta^2} - \epsilon \left[\omega \frac{dx}{d\theta} - \frac{\omega^3}{3} \left(\frac{dx}{d\theta} \right)^3 \right] + x = 0.$$

---

* See Appendix I for a proof of the existence of these solutions for ϵ sufficiently small.

We assume for x and ω the following power series in ϵ:

(5.2) $$x = x_0 + \epsilon x_1 + \epsilon^2 x_2 + \cdots$$

(5.3) $$\omega = \omega_0 + \epsilon \omega_1 + \epsilon^2 \omega_2 + \cdots .$$

We may now assume that the solution $x(\theta)$ of (5.1) has the period 2π; we also require that $dx/d\theta = 0$ for $\theta = 0$, i.e., that the velocity be zero at the time $t = 0$. Hence we require that the functions $x_i(\theta)$ all have the period 2π (this will serve to determine the constants ω_i), and that $dx_i/d\theta \,|_{\theta=0} = 0$.

Insertion of (5.2) and (5.3) into (5.1) yields

(5.4)
$$[x_0'' + \epsilon x_1'' + \cdots][\omega_0^2 + 2\epsilon\omega_0\omega_1 + \cdots]$$
$$- \epsilon[\omega_0 x_0' + \epsilon(\omega_0 x_1' + \omega_1 x_0') + \cdots]$$
$$+ \frac{\epsilon}{3}[\omega_0^3 + 3\omega_0^2\omega_1\epsilon + \cdots][x_0'^3 + 3\epsilon x_0'^2 x_1' + \cdots]$$
$$+ [x_0 + \epsilon x_1 + \epsilon^2 x_2 + \cdots] = 0.$$

The coefficient of the term of order zero in (5.4) is

(5.5) $$\omega_0^2 x_0'' + x_0 = 0,$$

with the conditions

(5.6) $$x_0(\theta) = x_0(\theta + 2\pi), \qquad x_0'(0) = 0.$$

The general solution of (5.5) is

$$x_0 = A_0 \cos \frac{\theta}{\omega_0} + B_0 \sin \frac{\theta}{\omega_0}.$$

If (5.6) is to be satisfied, it is necessary to take $\omega_0 = 1$ and $B_0 = 0$. The value $\omega_0 = 1$ was to be expected since the period of the oscillation for small ϵ (cf. (5.3)) should be in the neighborhood of 2π. Hence we have for x_0 and ω_0 finally

(5.7) $$x_0 = A_0 \cos \theta,$$

and

(5.8) $$\omega_0 = 1.$$

However, A_0 has not yet been determined. It will be fixed in the next step.

The first order term of (5.4) yields the differential equation for x_1 :

$$x_1'' + x_1 = -2\omega_1 x_0'' + x_0' - \tfrac{1}{3} x_0'^3$$

$$(5.9) \qquad = 2A_0\omega_1 \cos \theta - A_0 \sin \theta + \tfrac{1}{3} A_0^3 \sin^3 \theta$$

$$= 2A_0 \omega_1 \cos \theta + \left(\frac{A_0^3}{4} - A_0\right) \sin \theta - \frac{A_0^3}{12} \sin 3\theta.$$

To insure periodicity of $x_1(\theta)$ the resonance case must be avoided; hence we must impose the following conditions:

$$(5.10) \qquad \begin{cases} \dfrac{A_0^3}{4} - A_0 = 0, \\[2mm] \qquad \omega_1 = 0. \end{cases}$$

Thus, in particular, the period of the oscillation remains that of the free linear oscillation within second order terms in ϵ. We reject, naturally, $A_0 = 0$ as a solution of the first equation of (5.10) and also choose $A_0 = +2$. To take $A_0 = -2$ would lead only to a solution of different phase. For x_1 we have as general solution of (5.9)

$$(5.11) \qquad x_1 = A_1 \cos \theta + B_1 \sin \theta + \frac{1}{12} \sin 3\theta.$$

The constant B_1 is fixed by the requirement that $x_1'(0) = 0$. The constants A_1 and ω_2 are to be determined in the next step of the general process; it should be noted that the A_i will be *uniquely* determined once A_0 has been chosen. The first approximation to the periodic solution thus yields

$$(5.12) \qquad \begin{aligned} x &= 2 \cos \theta + \cdots \\ \omega &= 1 + \cdots . \end{aligned}$$

Accordingly, the amplitude of the oscillation is in first approximation a constant that is independent of the parameter ϵ and the period is 2π (i.e., that of the simple harmonic oscillation for $\epsilon = 0$) within second order terms in ϵ.

We have here an interesting example of a *bifurcation phenomenon*: For $\epsilon = 0$ the differential equation (5.1) has simple harmonic solutions which have arbitrary amplitudes but the same frequency for each given value of ω. For $\epsilon \neq 0$ we have seen that the nonlinear periodic solutions "bifurcate" from the simple harmonic solution which has a particular amplitude (i.e. the amplitude 2) and a particular frequency (corresponding to $\omega = 1$).

6. Relaxation oscillations*

We have already noted the change in the character of the solutions of (4.1) and consequently of (3.7) with increase of ϵ. In this section we are interested in discussing what occurs for large values of ϵ, and in particular what sort of limit cycle can be expected as $\epsilon \to \infty$ in (4.1). Oscillations of this kind (which appear to be of very widespread occurrence in nature) were given the name relaxation oscillations by van der Pol presumably because they exhibit two

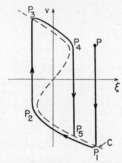

FIG. 6.1. Limit case of relaxation oscillations for $\epsilon \to \infty$.

distinct and characteristic phases: one during which energy is stored up slowly (in a spring or condenser) and another in which the energy is discharged nearly instantaneously when a certain critical threshold potential is attained.

In order to investigate the limit situation for $\epsilon \to \infty$, it is convenient to introduce a new independent variable $\xi = x/\epsilon$ in place of x. With this variable, equation (4.1) takes the form

$$(6.1) \qquad \frac{dv}{d\xi} = \epsilon^2 \frac{(v - v^3/3) - \xi}{v} = \epsilon^2 \frac{-F(v) - \xi}{v}.$$

The curve $\xi = -F(v)$ is plotted in Figure 6.1. The field direction is horizontal (i.e., $dv/d\xi = 0$) on the characteristic curve $\xi = -F(v)$

---

* For an interestingly written treatment of this subject see the paper of le Corbeiller [7].

for all ϵ, but at all other points in the plane the slope tends to become large without limit as $\epsilon \to \infty$, that is, the field directions would be nearly vertical at all points except those very near the characteristic curve C. With this in mind it is not very difficult to guess what the limit of the cycles will be (that is, rather, the limit of limit cycles for $\epsilon \to \infty$). Consider a solution curve of (6.1), for a large value of ϵ, which starts at P. The solution curve will be nearly a vertical straight line down to P_1, where it reaches C. At this point the slope is zero, but since the field direction at all points other than those near C is nearly vertical, the solution curve will tend to follow C, staying below it, until the vicinity of P_2 is reached. At this point the

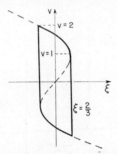

FIG. 6.2. Limit cycle with discontinuities for $\epsilon = \infty$.

curve turns almost vertically upwards until C is reached once more at P_3. The curve then follows C, staying above it, until P_4 is reached, where it turns vertically downwards again. This reasoning makes it highly plausible* that the limit of the limit cycles as $\epsilon \to \infty$ will be as shown in Figure 6.2. One might also argue directly from (6.1) that the limit cycle for $\epsilon \to \infty$ should consist of portions of the characteristic curve plus vertical straight lines, i.e., of either

$$(6.2) \qquad\qquad \xi = \text{constant},$$

or

$$(6.3) \qquad\qquad \xi = v - v^3/3.$$

It is of interest to compute the period of the relaxation oscillation, assuming that it is given as indicated by Figure 6.2. This can be done

---

* A rigorous proof of this statement is included as Appendix IV in this book.

by evaluating a line integral taken over the limit cycle. We wish to compute the period of the relaxation oscillation in terms of the original variables; hence we interpret ξ as x_1/ϵ and v as dx_1/dt_1 in the notation used for equation (3.6). The period T_1 of x_1 in terms of t_1 is given by

$$(6.4) \qquad\qquad T_1 = \epsilon \oint \frac{d\xi}{v}$$

since $dt_1 = dx_1/v = \epsilon d\xi/v$. Since the vertical (straight line) portions

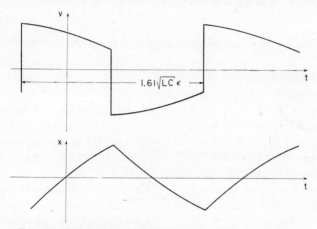

FIG. 6.3. Relaxation oscillations shown as functions of t.

of the cycle make no contribution to T_1 ($d\xi = 0$) we have in view of (6.3)

$$(6.5) \qquad T_1 = 2\epsilon \int_2^1 \frac{d(v - v^3/3)}{v} = 2\epsilon(\log v - v^2/2)\Big|_2^1$$

$$= 1.614\epsilon.$$

From (3.4) we have, finally, as the period $T = \sqrt{LC}\,T_1$ for the relaxation oscillation (for large $\epsilon = \alpha/\sqrt{LC}$) referred to the units of equation (3.3)

$$(6.6) \qquad\qquad T = 1.614\sqrt{LC}\,\epsilon = 1.614\,\alpha.$$

This relation, we repeat, yields a good approximation for T only for large ϵ. For small ϵ the period, as we have seen in the preceding section, is nearly independent of ϵ and is given approximately by

$$(6.7) \qquad\qquad T = 2\pi\sqrt{LC}.$$

As we see, the period of the oscillation depends upon the inductance and capacitance of the linear grid circuit in the latter case, but depends only upon the slope of the characteristic of the nonlinear element in the case of the relaxation oscillations.

The nature of the curves for v and $\dot{x}$ as functions of t is indicated in Figure 6.3 for ϵ large. Relaxation oscillations of the type shown in this figure are evidently to be expected in a triode oscillator, as we have remarked earlier, if the capacity C of the condenser is made small ($\epsilon = \alpha/\sqrt{LC}$), in particular if the condenser is suppressed.

7. *Higher order approximations for relaxation oscillations*

In the preceding section we have obtained an approximate formula for the period of relaxation oscillations of the van der Pol equation based on a geometrical discussion in the x, v-plane. Actually, the formula (6.6) gives the correct term of highest order in the asymptotic development for large values of ϵ of the period of the oscillation, as one might infer from the fact that it can be proved rigorously (cf. Appendix IV) that the limit cycles for $\epsilon \to \infty$ behave in the manner assumed in deriving the formula. However, the formula is not very accurate for values of ϵ which nevertheless are such that the oscillation should be regarded as a true relaxation oscillation; for example if $\epsilon = 10$ one sees from Figure 4.3 that the oscillation already exhibits the characteristic jerky motion expected of relaxation oscillations, but we shall see a little later that the period as given by the formula of the preceding section would be about 20 percent too low in this case. It is therefore important to improve the accuracy of the asymptotic formula for the period.

The problem of obtaining asymptotic developments for all quantities involved in relaxation oscillations has been solved completely and very generally by J. Haag [14], [15] in a series of important papers. Haag considers differential equations of very general type, which include the van der Pol equation as a special case. More recently, the complete asymptotic developments for the special case of the van der Pol equation have been derived by A. A. Dorodnitsyn [8]; this work is much more accessible than the work of Haag, perhaps because the latter takes a very general case. However, even the work of Dorodnitsyn is too extensive to be reproduced here, and we content ourselves with giving his formula for the period T_1 based on equation (3.6):

$$(7.1) \qquad T_1 = 1.614\epsilon + 7.014\epsilon^{-1/3} - \frac{22}{9} \frac{\log \epsilon}{\epsilon} + 0.0087\epsilon^{-1} + 0(\epsilon^{-4/3})$$

in which $0(\epsilon^{-4/3})$ represents a function which tends to zero of order $\epsilon^{-4/3}$. As we observe, the term of highest order is the same as was given by (6.5) above. However, the term of next highest order dies out only like $\epsilon^{-1/3}$ so that it makes a relatively large contribution even for such a value as $\epsilon = 10$—in fact, as we remarked above, the error is of the order of 20 percent in this case if only the first term in the development is used.

Because of the importance of the study of the asymptotic behavior of relaxation oscillations in general, it would seem worth while to indicate how one can carry out such developments in a case which, unlike that of the van der Pol equation, is simple enough to be treated explicitly. At the same time one obtains some insight into the peculiar difficulties to be expected in treating other cases. We consider for this purpose the differential equation

$$(7.2) \qquad \ddot{x} + \epsilon F(\dot{x}) + x = 0,$$

with $F(v)$ the piece-wise linear function indicated in Figure 7.1 and defined by

$$(7.3) \qquad \begin{cases} F(v) = \pm 2 - v, & |v| \geq 1 \\ F(v) = v, & |v| \leq 1. \end{cases}$$

With this function $F(v)$ it is clear that the oscillations are of the self-sustained type; in fact, this characteristic could be considered as a rough approximation to the characteristic of the van der Pol equation. Actually, Haag [15] considers this case also, but treats it by specializing his general formulas.

As in the preceding section, there is an advantage in introducing a new variable x_1 replacing x by the relation

$$(7.4) \qquad x_1 = \frac{1}{\epsilon} x,$$

in terms of which (7.2) can be written

$$(7.5) \qquad \frac{dv}{dx_1} = \epsilon^2 \frac{-F(v) - x_1}{v},$$

with v defined by

$$(7.6) \qquad v = \frac{dx}{dt} = \epsilon \frac{dx_1}{dt}.$$

By the same arguments as in the preceding section, the limit of limit cycles of (7.5) as $\epsilon \to \infty$ would be the two vertical segments between $(1, 1)$ and $(1, -3)$ and $(-1, -1)$ and $(-1, 3)$ (cf. Figure 7.1) and the segments of the characteristics between the appropriate pairs of

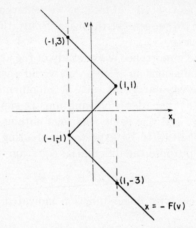

FIG. 7.1. A piece-wise linear characteristic.

these points. Also, since $dt = dx/v$, we have the following approximation for the period T_1 of the relaxation oscillation:

$$(7.7) \qquad T_1 = \oint \frac{dx}{v} \simeq 2\epsilon \int_{-1}^{1} \frac{dx_1}{v} = 2\epsilon \int_{-1}^{1} \frac{dx_1}{2 - x_1} = 2\epsilon \log 3,$$

since $dx = 0$ on the vertical parts of the cycle. We shall see that this does in fact yield the highest order term in the asymptotic development of T_1.

Instead of seeking the closed solution curve of (7.5) it is somewhat more convenient to work with the equivalent second order differential equations since they have constant coefficients. These equations are

$$(7.8) \qquad \ddot{x}_1 - (\epsilon \dot{x}_1) + x_1 = 0 \qquad \text{for } |\epsilon \dot{x}_1| = |v| \leq 1, \qquad \text{and}$$

$$(7.9) \qquad \ddot{x}_1 - (2 - \epsilon \dot{x}_1) + x_1 = 0 \qquad \text{for } |\epsilon \dot{x}_1| = v \geq 1,$$

as one sees from (7.2) and (7.4). What is to be done is rather obvious: The solutions $(x_1(t), v(t))$ of (7.8) and (7.9) must be pieced together across the lines $v = +1$ and $v = -1$ in the x_1, v-plane, and the initial conditions must be chosen so that the solution (x_1, v) will form a closed curve in the x_1, v-plane. Once these conditions have been written down, they can be developed asymptotically for large values of ϵ.

The general solutions of (7.8) and (7.9) are

(7.10) $x_1^{(1)} = c_1 e^{\lambda_1 t} + c_2 e^{\lambda_2 t}$ for $|v| \le 1$,

(7.11) $x_1^{(2)} = 2 + c_3 e^{-\lambda_1 t} + c_4 e^{-\lambda_2 t}$ for $v \ge 1$,

with

(7.12) $\lambda_1 = \dfrac{\epsilon + \sqrt{\epsilon^2 - 4}}{2}, \qquad \lambda_2 = \dfrac{\epsilon - \sqrt{\epsilon^2 - 4}}{2}.$

The c_i are, of course, arbitrary constants. As initial conditions at the time $t = 0$ we take $x_1 = 1 + \delta$, $v = \epsilon \dot{x}_1 = 1$ for both $x_1^{(1)}$ and $x_1^{(2)}$—in other words, we begin at the point $(1 + \delta, 1)$ in the x_1, v-plane (cf. Figure 7.2)—and afterwards determine $\delta(\epsilon)$ in such a way that a

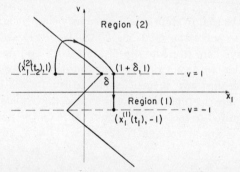

Fig. 7.2. Integral curve for case of a piece-wise linear characteristic.

periodic solution results. The periodicity condition will be satisfied (again cf. Figure 7.2) if $x_1^{(2)}(t_2) = -x_1^{(1)}(t_1)$, on account of the symmetry of the problem, with t_1 and t_2 to be determined from $v^{(1)}(t_1) = -1$ and $v^{(2)}(t_2) = +1$.

In the region (1) in which $|v| \le 1$ we have therefore the following conditions on c_1 and c_2:

(7.13) $1 + \delta = c_1 + c_2$,

(7.14) $1 = \epsilon(\lambda_1 c_1 + \lambda_2 c_2).$

For the periodicity conditions we shall need later the relations

$$(7.15) \qquad x_1^{(1)}(t_1) = c_1 e^{\lambda_1 t_1} + c_2 e^{\lambda_2 t_1}, \qquad\qquad \text{and}$$

$$(7.16) \qquad -1 = \epsilon(\lambda_1 c_1 e^{\lambda_1 t_1} + \lambda_2 c_2 e^{\lambda_2 t_1}),$$

with c_1 and c_2 defined by (7.13) and (7.14). Similarly in the region (2) we have for c_3 and c_4 the conditions

$$(7.17) \qquad\qquad 1 + \delta = 2 + c_3 + c_4, \qquad\qquad \text{and}$$

$$(7.18) \qquad\qquad 1 = \epsilon(-\lambda_1 c_3 - \lambda_2 c_4),$$

while the equations to be used for the periodicity conditions are

$$(7.19) \qquad x_1^{(2)}(t_2) = 2 + c_3 e^{-\lambda_1 t_2} + c_4 e^{-\lambda_2 t_2}, \qquad\qquad \text{and}$$

$$(7.20) \qquad\qquad 1 = \epsilon(-\lambda_1 c_3 e^{-\lambda_1 t_2} - \lambda_2 c_4 e^{-\lambda_2 t_2}).$$

The periodicity conditions are furnished by (7.16) and (7.20) together with the condition

$$(7.21) \qquad\qquad x_1^{(2)}(t_2) = -x_1^{(1)}(t_1).$$

One observes that even in this particularly simple case the calculations needed to solve the relations (7.13) to (7.21) are not entirely trivial. It pays here to make use of knowledge and insights already gained in order to facilitate the asymptotic development—a not uncommon circumstance in general when asymptotic developments are wanted. To this end we consider first the equation (7.20) for the time t_2 required to pass in the region (2) from the point $(1 + \delta, 1)$ to the point $(x_2^{(2)}(t_2), 1)$: From (7.12) we see that λ_1 and λ_2 have the developments

$$(7.22) \qquad \begin{cases} \lambda_1 = \epsilon\left(1 - \dfrac{1}{\epsilon^2} - \dfrac{1}{\epsilon^4} - \dfrac{2}{\epsilon^6} - \dfrac{5}{\epsilon^8} - \dfrac{14}{\epsilon^{10}} + \cdots\right), \\[2mm] \lambda_2 = \dfrac{1}{\epsilon}\left(1 + \dfrac{1}{\epsilon^2} + \dfrac{2}{\epsilon^4} + \dfrac{5}{\epsilon^6} + \dfrac{14}{\epsilon^8} + \cdots\right). \end{cases}$$

From the discussion which led to (7.7) we know that t_2 must be of order ϵ for ϵ large; in addition t_2 must be negative since points on the solutions of (7.5) move in the direction of the arrows (cf. Figure 7.2) as t increases, and the point $(1 + \delta, 1)$ corresponds to $t = 0$. It follows that $e^{-\lambda_1 t_2}$ behaves like $e^{+\epsilon^2}$ while $e^{-\lambda_2 t_2}$ is finite for ϵ large.

It follows therefore that $\lambda_1 c_3$ must tend to zero like $e^{-\epsilon^2}$ in order to make relation (7.20) valid. From (7.18) we may therefore conclude that

$$(7.23) \qquad c_4 = -\frac{1}{\epsilon \lambda_2} = -\frac{\lambda_1}{\epsilon} = -1 + \frac{1}{\epsilon^2} + \frac{1}{\epsilon^4} + \frac{2}{\epsilon^6} + \frac{5}{\epsilon^8} + \cdots$$

aside from terms of exponential order, which we neglect.* From (7.17) we then obtain at once the expansion for δ, as follows:

$$(7.24) \qquad \delta = \frac{1}{\epsilon^2} + \frac{1}{\epsilon^4} + \frac{2}{\epsilon^6} + \frac{5}{\epsilon^8} + \cdots.$$

It is rather strange that we are thus able to determine asymptotically a point on the limit cycle (at least within terms of exponential order) without using the periodicity condition (7.21) explicitly; however, to make up for this, we still have not fixed the development of the term $c_3 e^{-\lambda_1 t_2}$, which is not necessarily of exponential order, and we shall have to make use of (7.20) again for this purpose later on. It is also worth noting that $\delta \simeq 1/100$ for $\epsilon = 10$, which as we see from Figure 7.2 indicates that the convergence toward the limit of limit cycles for $\epsilon = \infty$ is rather rapid. From (7.13) and (7.14) we can now obtain the developments for c_1 and c_2 :

$$(7.25) \qquad \begin{cases} c_1 = -\dfrac{2}{\epsilon^4} - \dfrac{8}{\epsilon^6} - \dfrac{30}{\epsilon^8} - \cdots, \\[2mm] c_2 = 1 + \dfrac{1}{\epsilon^2} + \dfrac{3}{\epsilon^4} + \dfrac{10}{\epsilon^6} + \dfrac{35}{\epsilon^8} + \cdots. \end{cases}$$

It is convenient next to obtain the development for t_1 from (7.16) written in the form

$$(7.26) \qquad e^{\lambda_1 t_1} = -\frac{1}{\epsilon \lambda_1} - \frac{\lambda_2}{\lambda_1} c_2 e^{\lambda_2 t_1}.$$

Since λ_2 is of order $1/\epsilon$ and we expect that $t_1 \to 0$ as $\epsilon \to \infty$—once more on the basis of the discussion which led to (7.7)—we can obtain the lowest order term in the development for t_1 by setting $e^{\lambda_2 t_1} = 1$, then reinserting this value for t_1 in the right-hand side of (7.26) to

---

* From this the character of our developments as asymptotic rather than convergent is clear.

obtain two terms in the development for t_1, and so on; the result of such a calculation by iterations is

$$(7.27) \quad t_1 = \frac{1}{\epsilon}\left[2\log\epsilon + \frac{3\log\epsilon}{\epsilon^2} - \frac{2}{\epsilon^2} + \frac{1}{2}\frac{\log^2\epsilon}{\epsilon^4} + 0\left(\frac{\log\epsilon}{\epsilon^4}\right)\right].$$

We observe that t_1 does tend to zero as $\epsilon \to \infty$, as we expected, but only like $(\log \epsilon)/\epsilon$. We also observe that the development is rather complicated, including as it does terms in $(\log \epsilon)/\epsilon$, $(\log^2 \epsilon)/\epsilon^4$, etc. As next step we take the sum of (7.15) and (7.19), make use of the periodicity condition (7.21), and eliminate the terms $c_2 e^{\lambda_2 t_1}$ and $c_3 e^{-\lambda_1 t_2}$ by using (7.16) and (7.20)* to obtain finally the following equation for t_2 :

$$(7.28) \qquad 1 + c_1\left(1 - \frac{\lambda_1}{\lambda_2}\right)e^{\lambda_1 t_1} = c_4\left(\frac{\lambda_2}{\lambda_1} - 1\right)e^{-\lambda_2 t_2}.$$

We have now made use of all of the relations (7.13) to (7.21) inclusive. From (7.28) the following development for t_2 is obtained without difficulty by making use of the development for t_1 given in (7.27):

$$(7.29) \quad t_2 = -\epsilon\left[\log 3 + \frac{2}{3}\frac{\log\epsilon}{\epsilon^2} + \left(\frac{4}{3} - \log 3\right)\frac{1}{\epsilon^2} + \frac{4}{9}\frac{\log^2\epsilon}{\epsilon^4} + 0\left(\frac{\log\epsilon}{\epsilon^4}\right)\right].$$

Since the period T_1 of the relaxation oscillation is given by $T_1 = 2(t_1 - t_2)$ we obtain finally**

$$(7.30) \quad T_1 = 2\left[\epsilon\log 3 + \frac{8}{3}\frac{\log\epsilon}{\epsilon} + \left(\frac{4}{3} - \log 3\right)\frac{1}{\epsilon} + \frac{4}{9}\frac{\log^2\epsilon}{\epsilon^3}\right] + 0\left(\frac{\log\epsilon}{\epsilon^3}\right).$$

On comparison with (7.7) we see that the highest order term in the asymptotic development is indeed given by (7.7). It is clear that asymptotic developments for the limit cycle itself could also be given without difficulty.

* Here we use (7.20) once more to get rid of the troublesome term $c_3 e^{-\lambda_1 t_2}$.

** This development coincides with that given by Haag [15] except for the coefficient of the term in $(\log \epsilon)/\epsilon$, which appears to be given incorrectly by Haag.

It is of interest to compare the values given by (7.30) with values obtained by computations based on the exact solution of the differential equation, which can be obtained without too much difficulty in the present case. For $\epsilon = 10$ the asymptotic series for T_1 given by (7.30) was found in this way to be in error by less than 0.1 percent, and even for $\epsilon = 5$ the error was less than 0.8 percent. The first four terms in the series, as given in (7.30), were used in making the computation.

The development (7.1) for the period T_1 in the case of the van der Pol equation should be compared with (7.30). We observe that in both cases the terms of highest order are linear in ϵ, but the terms of next higher order are of different orders in the two cases: of order $\epsilon^{-1/3}$ in the case of the van der Pol equation and of order $\epsilon^{-1} \log \epsilon$ in the case studied here. This indicates a rather strong sensitivity to the shape of the characteristic. Another observation should be made: the various quantities considered behave differently on different portions of the limit cycle as regards their order of magnitude in ϵ, so that these various portions must be investigated separately in general, and this is one of the principal reasons for the complexity of the developments in the papers of Haag and Dorodnitsyn.

B. Forced Oscillations in Self-sustained Systems

8. A typical physical problem

This part of the present chapter is devoted to a study of the oscillations which occur when a periodic excitation is applied to a system whose free oscillations are of the self-sustained type. A typical and important case is the electrical system indicated in Figure 8.1, which leads to the differential equation first treated by van der Pol [32]. This vacuum tube circuit differs from the circuit discussed in Section 1 above (cf. Figure 1.1)* in that the oscillatory component is in the grid circuit rather than the anode circuit and, in addition, a source $E = P_0 \sin \omega_1 t$ of alternating voltage is present

---

* By adding a periodic excitation to the circuit of Figure 1.1 in an appropriate way one would, of course, obtain a system of the type under consideration here, which would also lead to the same general results. The circuit of Figure 8.1 was chosen because it yields the desired differential equation in a straightforward way.

in the grid circuit, as indicated in the figure. As before, the grid and plate circuits are inductively coupled. The differential equations for the system in terms of the current i in the grid circuit, the current i_a in the anode circuit, and the grid potential u_g are readily derived; they are:

(8.1)
$$\begin{cases} L\dfrac{di}{dt} + Ri + u_g - M\dfrac{di_a}{dt} = P_0 \sin \omega_1 t, \\[2mm] \qquad\qquad C\dfrac{du_g}{dt} = i. \end{cases}$$

We have ignored the current in the grid itself in deriving (8.1). We now make the assumption that the anode current i_a depends

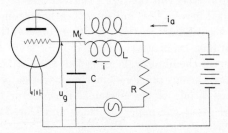

Fig. 8.1. A self-excited system with periodic excitation.

only upon the potential drop u_g between the grid and the filament* and that the relation between these two quantities is as follows:

(8.2)
$$i_a = Su_g\left(1 - \frac{u_g^2}{3K^2}\right),$$

with S and K positive constants. The quantity S is sometimes called the *steepness* of the characteristic and K is called the *saturation potential*. We can replace both i and i_a in (8.1) in terms of u_g through use of (8.2). It is, however, convenient to introduce first the following new quantities:

(8.3)
$$v = \frac{u_g}{K}, \qquad \alpha = \frac{MS - RC}{LC}, \qquad \gamma = \frac{1}{3}\frac{MS}{LC},$$
$$B = \frac{P_0}{K}, \qquad \omega_0^2 = \frac{1}{LC},$$

---

* This condition is satisfied closely if the amplification factor of the tube is large (cf. the discussion in Section 1).

in terms of which (8.1) becomes

$$(8.4) \qquad \frac{d^2 v}{dt^2} - \alpha \frac{dv}{dt} + \gamma \frac{d}{dt}\,(v^3) + \omega_0^2 v = B\omega_0^2 \sin \omega_1 t.$$

It is always assumed that α is positive, in order that self-excited oscillations can occur. We note that ω_0 is the frequency of the free linear vibration of the oscillatory circuit. The differential equation (8.4) will be the basis for the discussion in the remainder of this chapter. The left-hand side of equation (8.4) is essentially the same as the left-hand side of equation (1.17) above; in other words, the free oscillations in the present case are not essentially different from those treated earlier in Section 1 and Section 3 of this chapter; practically the only difference is that the potential drop rather than the current is taken as the basic quantity here. Also, the theory for the equation $\ddot{x} - \alpha \dot{x} + \gamma \dot{x}^3 + \omega_0^2 x = F \cos \omega_1 t$ is the same as for (8.4) since (8.4) results by differentiating the latter equation with respect to t and identifying v with $\dot{x}$ and $-F\omega_1$ with $B\omega_0^2$.

9. The method of van der Pol for the forced oscillations

The periodic solutions of (8.4) were investigated first by van der Pol [32], who devised a method of attacking the equation which is different from any others used hitherto in this book. Many writers on nonlinear vibrations, the Russian writers particularly, prefer to make use of the method of van der Pol to discuss forced oscillations of all systems, including those with nonlinear restoring forces. The author is inclined to feel, however, that the method of van der Pol has no advantages over the simpler and more direct perturbation or iteration methods used in the cases treated up to now. Even in the present case the perturbation method* is more straightforward than van der Pol's method if one is simply interested in obtaining the response curves—that is, the curves showing the relation between the amplitude and frequency of the harmonic oscillations. However, the treatment of the *stability* of these oscillations is greatly facilitated by using van der Pol's method in the manner devised by Andronow and Witt [1], which will be treated in the next section. Also, the method of van der Pol can be used to study certain oscillations which are

---

* In Appendix I the existence of the perturbation series for the present case (as well as others) is proved, and the response curves are derived.

not periodic functions, but rather almost-periodic functions of the time. In fact, these results of van der Pol are among the most beautiful and striking in the whole range of problems treated in this book.

Van der Pol's method is an approximate method of solving the equation (8.4) which is based—like the perturbation method—on the assumption that the coefficients of the damping terms and the amplitude of the excitation are small so that the motion is not far from that of the linear undamped free vibration; thus the constants α, γ, and B in (8.4) are assumed to be small quantities of the same order of magnitude. In terms of the physical problem these conditions could be satisfied by taking the steepness of the characteristic and the amplitude of the excitation small enough. In addition, it is assumed that ω_1, the frequency of the excitation, differs from ω_0, the frequency of the free oscillation, by an amount which is also a small quantity of the same order of magnitude as α, γ, and B.

The essential step in van der Pol's procedure, however, consists in taking for $v(t)$ a solution in the form

$$(9.1) \qquad v(t) = b_1(t) \sin \omega_1 t + b_2(t) \cos \omega_1 t$$

in which the functions $b_i(t)$ are assumed to be "slowly varying functions of the time," or, in other words, that the motion is essentially an oscillation with the frequency ω_1 of the excitation but with slowly varying amplitude and phase. This latter assumption is to be interpreted as meaning that the quantities $\dot{b}_i$ and $\ddot{b}_i$ are small of first and second order respectively in α (and γ and B).* Finally, it is implicitly assumed in writing the solution in the form given by equation (9.1) that possible terms involving higher harmonics of $\omega_1 t$ are of order α or higher.

By differentiation of (9.1) we find

$$(9.2) \quad \dot{v} = \dot{b}_1 \sin \omega_1 t + \dot{b}_2 \cos \omega_1 t + b_1 \omega_1 \cos \omega_1 t - b_2 \omega_1 \sin \omega_1 t,$$

$$(9.3) \quad \ddot{v} = 2\dot{b}_1 \omega_1 \cos \omega_1 t - 2\dot{b}_2 \omega_1 \sin \omega_1 t - b_1 \omega_1^2 \sin \omega_1 t$$

$$- b_2 \omega_1^2 \cos \omega_1 t + \ddot{b}_1 \sin \omega_1 t + \ddot{b}_2 \cos \omega_1 t.$$

* It would be possible to set up a formal perturbation scheme in which the quantities $\dot{b}_i$ and $\ddot{b}_i$ would have automatically the desired order of magnitude. Such a development would probably not be convergent, but would have asymptotic validity.

We must also calculate the quantity v^3 which occurs in (8.4); in doing so we use the familiar identity $\sin^3 \theta = \frac{3}{4} \sin \theta - \frac{1}{4} \sin 3\theta$ and the corresponding identity for $\cos^3 \theta$ to obtain the relation

(9.4) $\qquad v^3 = \frac{3}{4}(b_1^2 + b_2^2)(b_1 \sin \omega_1 t + b_2 \cos \omega_1 t) + \cdots .$

The dots refer to terms of frequency $3\omega_1$. We now insert $\dot{v}$, $\ddot{v}$, and v^3 as given by (9.2), (9.3), (9.4) in (8.4), reject all terms of order higher than the first, bearing in mind that the quantities $\dot{b}_i$ are of first order and the quantities $\ddot{b}_i$ of second order. By equating the coefficients of $\cos \omega_1 t$ and $\sin \omega_1 t$ in the two sides of equation (8.4) we obtain the differential equations

(9.5)
$$
\begin{cases}
2\dot{b}_1 + \dfrac{\omega_0^2 - \omega_1^2}{\omega_1} b_2 - \alpha b_1 \left(1 - \dfrac{b^2}{a_0^2} \right) = 0 \\[3mm]
2\dot{b}_2 + \dfrac{\omega_1^2 - \omega_0^2}{\omega_1} b_1 - \alpha b_2 \left(1 - \dfrac{b^2}{a_0^2} \right) = -B \dfrac{\omega_0^2}{\omega_1},
\end{cases}
$$

in which we have introduced the quantities

(9.6) $\qquad a_0^2 = \dfrac{\alpha}{\frac{3}{4}\gamma}, \qquad b^2 = b_1^2 + b_2^2.$

The equations (9.5), which were first derived by van der Pol, form the basis for all of the discussion in the remainder of this chapter. The important quantity b is the "amplitude" of the vibration, as we see from (9.1); the quantity a_0 is the amplitude, within terms of lowest order in α, of the free nonlinear vibration of frequency ω_0, as we shall verify shortly. We now introduce in addition the quantity Δ given by

(9.7) $\qquad\qquad\qquad \Delta = 2(\omega_0 - \omega_1),$

which measures the difference between the frequencies of the free oscillation and the excitation; this quantity, or others proportional to it, will be referred to on occasion as the *detuning*. Since Δ is assumed to be a quantity of first order, it follows at once from the fact that $\omega_0^2 - \omega_1^2 = (\omega_0 - \omega_1)(\omega_0 + \omega_1) = \omega_1 \Delta$ within terms of first order, that the equations (9.5) can be written in the following form:

(9.8)
$$
\begin{cases}
2\dot{b}_1 + b_2 \Delta - \alpha b_1 \left(1 - \dfrac{b^2}{a_0^2} \right) = 0 \\[3mm]
2\dot{b}_2 - b_1 \Delta - \alpha b_2 \left(1 - \dfrac{b^2}{a_0^2} \right) = -B\omega_0 .
\end{cases}
$$

As in all of our previous discussions of forced oscillations we are interested at the outset primarily in periodic solutions of the differential equation (8.4) which have the frequency ω_1 of the excitation, that is, in solutions which we are accustomed to calling the harmonic oscillations. In order to obtain them it is necessary, in view of the form (9.1) of the solution assumed here, to require that b_1 and b_2 be constants. If this is done, the equations (9.8) become algebraic relations which determine the amplitude b of the forced oscillation in terms of Δ, the detuning, with the amplitude B of the excitation as a parameter. In other words, the response curves can be obtained directly from (9.8). In particular, if we take $B = 0$, i.e. if we consider the *free oscillation*, we obtain easily from (9.8) the result $a_0 = b$, $\Delta = 0$, since b_1 and b_2 cannot both vanish and $\alpha \neq 0$; this verifies the statement made earlier that a_0 is the amplitude of the free nonlinear oscillation of frequency ω_0 .

In his basic paper van der Pol [32] proceeded to analyze the stability of the periodic solutions obtained by the method just explained. In doing so, he replaced b_1 and b_2 in (9.8) by $b_1 + \delta b_1$ and $b_2 + \delta b_2$, obtained linear differential equations with constant coefficients for δb_1 and δb_2 , and determined the stability of the periodic solutions in accordance with the behavior of the solutions of the latter differential equations in the obvious manner. Instead of studying the stability of the periodic solutions in this way we prefer rather to follow the method developed by Andronow and Witt [1]. *The basic idea of the method of Andronow and Witt is to study the solutions of the differential equations* (9.8) *quite generally*, rather than to confine attention to the special solutions $b_1 = $ constant, $b_2 = $ constant; and this is made feasible by the fact that the two equations (9.8) are equivalent to the single equation of first order $db_1/db_2 = P(b_1 , b_2)/Q(b_1 , b_2)$, which in turn can be treated by the methods of Chapter III. In fact, this work of Andronow and Witt furnishes, among other things, a beautiful application of the theory of the singularities of first order differential equations which was worked out in detail in Chapter III: in particular, we see at once that the harmonic solutions of the original differential equation (8.4) are correlated with the singularities of the first order system (9.8).

In the next section we shall discuss the method of Andronow and Witt in a general way, followed in Section 11 by the derivation of the response curves for the harmonic oscillations, although the latter could also have been derived here simply on the basis of equations (9.8). In Section 12 the stability of the harmonic oscillations is treated in detail.

10. The method of Andronow and Witt

As indicated at the end of the preceding section, our object is to study in some detail the solutions of equations (9.8), which yield the amplitude and phase of the forced oscillations of the van der Pol equation. For this purpose it is useful to introduce still another set of new variables and parameters, as follows:

$$(10.1) \qquad x = b_1/a_0, \qquad y = b_2/a_0, \qquad r^2 = x^2 + y^2, \qquad \tau = t\alpha/2,$$

$$(10.2) \qquad \sigma = \Delta/\alpha, \qquad F = -B\omega_0/a_0\alpha.$$

It is important to observe that the new independent variable τ has the same sign as t since α is always taken positive. The quantity r is the ratio of the amplitude of the forced oscillation to that of the free oscillation, the quantity σ represents the detuning, and F is proportional to the amplitude of the excitation. In terms of these quantities the equations (9.8) can be written in the form

$$(10.3) \qquad \begin{cases} \dfrac{dx}{d\tau} = -\sigma y + x(1 - r^2) \\[2mm] \dfrac{dy}{d\tau} = F + \sigma x + y(1 - r^2), \end{cases}$$

or, equivalently, in the form

$$(10.4) \qquad \frac{dy}{dx} = \frac{F + \sigma x + y(1 - r^2)}{-\sigma y + x(1 - r^2)} = \frac{P(x, y)}{Q(x, y)}.$$

In terms of the new variables the solutions of the form (9.1) of the original differential equation (8.4) take the form

$$(10.5) \qquad v/a_0 = x \sin \omega_1 t + y \cos \omega_1 t,$$

with x and y to be determined from (10.3). As we have already remarked, periodic solutions of frequency ω_1 of the original differential equation (8.4) are therefore to be correlated with solutions $x =$ constant, $y =$ constant of (10.3) and *hence with the singular points of*

(10.4), i.e. the points (x, y) at which $P(x, y)$ and $Q(x, y)$ vanish simultaneously. The distance from the origin to any such point represents the ratio b/a_0 of the amplitude of the forced oscillation to the amplitude of the free oscillation, as we see from (10.1) and (9.6). *The stability of any particular harmonic solution of (8.4) is then*, according to the basic idea of Andronow and Witt, *to be decided in accordance with the character of the singularity of (10.4) with which it is correlated.* For example, if the singularity of (10.4) were a saddle point, it is clear that the corresponding periodic solution (10.5) should be considered to be unstable, since a slight disturbance would result in general in large changes in x and y. On the other hand, if the singularity were a stable spiral point it follows that a slight disturbance would lead to a solution which would tend with increasing t to the periodic solution since $x(t)$ and $y(t)$ would tend to the constants characterizing the periodic solution with increasing t; hence such a solution should be considered to be stable. The general principle by which the decision as to the stability of a given oscillation is to be made is evidently the following: *Stable harmonic oscillations are correlated with the singularities of (10.4) whith are stable in the sense defined in Chapter III, and unstable harmonic oscillations are correlated with unstable singularities.*

Once the stability problem has been formulated in this way one can foresee another interesting possibility: *It may happen that the solution curves of (10.4) in the x, y-plane have a stable limit cycle toward which all solution curves tend*, just as in many of the cases of the free oscillations of self-excited systems treated earlier in this chapter. In such a case we know that $x(\tau)$ and $y(\tau)$ tend to periodic functions with the same period in τ; in other words x and y in (10.5) would tend to periodic functions having the same period and hence the solution of the original differential equation (8.4) would be one in which the amplitude and phase after the lapse of sufficient time vary slowly but periodically in the time, or, as one says, the oscillation is affected by amplitude and phase modulation. Later on we shall indicate how one can show that the functions $x(t)$ and $y(t)$ are approximately simple harmonic functions with the frequency $\omega_1 - \omega_0$ provided that σ and F are large enough; it follows that the solution (10.5) could be written in such a case as the sum of two harmonic oscillations with frequencies ω_0 and ω_1. In other words, the vibration would be a superposition of two simple harmonic oscillations, one with the frequency of the excitation, the other with the frequency of the free oscillation, and the latter oscillation would not be damped out. Such oscillations have often been observed in electrical circuits of the type considered here. That such a vibration might be possible in the present case, although it is not possible in the case of linear systems in which the free oscillation is always damped out, can be attributed to the fact that the "damping" force acts in the present cases in just such a way as to maintain the free oscillation.

In the next two sections we shall study the differential equations (10.3) and (10.4) in detail. First of all the response curves for the harmonic solutions of the original differential equation (8.4) will be obtained, after which the stability of these oscillations will be investigated. Finally, the existence and character of certain stable nonharmonic oscillations, which are correlated with limit cycles of equation (10.4), will be treated.

11. Response curves for the harmonic oscillations

We have already seen that the amplitude of any harmonic oscillation of the van der Pol equation is obtained by equating the right-hand sides of equations (10.3) to zero, which amounts to saying that the location of the singular points of the first order differential equation (10.4) is determined. If (x_0, y_0) is such a singular point, the following conditions hold:

$$(11.1) \quad \begin{cases} -\sigma y_0 + x_0(1 - r_0^2) = 0 \\ F + \sigma x_0 + y_0(1 - r_0^2) = 0 \end{cases} \quad \text{with } r_0^2 = x_0^2 + y_0^2 = \rho.$$

The quantity ρ represents the square of the ratio of the amplitudes of the harmonic oscillation and the free oscillation, and we repeat that F is proportional to the amplitude of the excitation while σ is proportional to the detuning, or the difference between the frequency of the harmonic oscillation and the free oscillation. By determining x_0 and y_0 in terms of F, σ, and ρ from (11.1) and then inserting these values in $x_0^2 + y_0^2 = \rho$ one finds readily the following relation between ρ, σ, and F:

$$(11.2) \quad \rho[\sigma^2 + (1 - \rho)^2] = F^2,$$

which was given first by van der Pol. For each frequency of the harmonic oscillation (i.e. for each value of σ) and for a given amplitude of the excitation (given by F) equation (11.2) furnishes through the values of ρ the corresponding amplitudes of the harmonic oscillations. Equation (11.2) thus yields what we call the response curves.

We proceed to discuss the curves given by equation (11.2) in the σ, ρ-plane for various values of the parameter F. Clearly we are interested only in positive values of ρ. Evidently the curves are symmetrical with respect to the ρ-axis. For $F = 0$ we have the case of the free oscillation; in this case we observe from (11.2) that $\sigma = 0$, $\rho = 1$, and $\rho = 0$, with σ arbitrary, are the only values of σ and ρ which are possible. Thus, as we have seen earlier, there is only one free oscillation of non-vanishing amplitude represented by the point $\sigma = 0$, $\rho = 1$. It is mainly this fact which causes the response curves in the present case to differ so much in appearance from those encountered in Chapter IV in dealing with the systems having nonlinear restoring forces, in which the free oscillations were not unique. If we consider next cases in which F is small, but different from zero, we expect ρ to be nearly unity or nearly zero so that the response curves would be ovals which are approximately ellipses $\sigma^2 + (1 - \rho)^2 = F^2$ with their centers at $\sigma = 0$, $\rho = 1$ and in addition branches running near the σ-axis. The ovals expand with increasing F. For F not too large, therefore, we expect three values of ρ for each not too large value of σ. However, it is found that the upper branches—the ovals—change from closed curves to open curves on passing a certain critical value of F, and that the response curve for this critical value of F has a double point on the ρ-axis with $\rho = 1/3$. For values of F larger than the critical value the response curves are all open curves with no double points, but ρ is still not a single-valued function of σ on these curves for values of F near the critical value and for not too large values of σ. Figure 11.1 shows a number of the response curves for various values of F; we repeat the highly important fact that *each point on any such curve yields the amplitude which is correlated with the frequency of a possible harmonic oscillation for a given amplitude of the excitation.* As we shall see shortly, the transition value of F at which the two distinct branches of the response curves coalesce is given by $F^2 = 4/27$. On Figure 11.1 a closed curve $E(\rho, \sigma) = 0$ is drawn; this curve is *an ellipse which is the locus of the vertical tangents of the response curves,* as we shall show in a moment. This curve plays an important role in the discussion to

follow.   As we shall see in the next section, the vertical tangencies of
the response curves which lie on the upper half of this ellipse represent
transitions from stable to unstable oscillations.   However, the
vertical tangencies on the lower half of the ellipse have no such

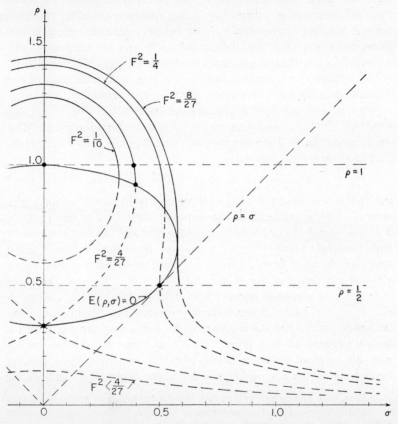

FIG. 11.1. Response curves for the harmonic oscillations of a self-sustained
system.

significance: the portion of any response curve which lies below the
vertical tangency on the upper half of the ellipse (assuming that the
curve cuts the ellipse at all) contains only unstable points—that is
to say, the singularities of (10.4) correlated with these points are
unstable.   In addition, all points below the line $\rho = 1/2$ are found

to be unstable, so that the response curves for sufficiently high values of F^2 have unstable lower portions although they have no vertical tangencies at all.* Unstable parts of the response curves are indicated by dashed lines. As indicated on Figure 11.1 the transition from response curves with vertical tangents to those without them occurs when $F^2 = 8/27$; the corresponding response curve is tangent to the ellipse at its vertex. Also indicated on the figure is the fact that the straight lines $\rho = \pm\sigma$ are tangent to the ellipse $E(\rho, \sigma) = 0$ at the points where $\rho = 1/2$, and the value of F^2 on the response curve through these points is $1/4$; these features of the response diagram have importance for the later discussion.

We now indicate briefly how one can verify the above statements regarding the geometric character of the response curves. The condition $d\sigma/d\rho = 0$ for the vertical tangents is readily found from (11.2) to lead to the relation:

$$(11.3) \qquad E(\rho, \sigma) = \sigma^2 + (1 - \rho)(1 - 3\rho) = 0,$$

and this is the equation of the ellipse discussed above which is the locus of vertical tangents of the response curves. For small values of F, the response curves have only one vertical tangent for $\sigma > 0$, while for larger values they have two; the transition obviously takes place, as we infer from Figure 11.1, for the response curve which contains the vertex of the ellipse at $\sigma = 0$, $\rho = 1/3$, from which one finds for this response curve $F^2 = 4/27$ from (11.2). The point $\sigma = 0$, $\rho = 1/3$ on the curve for $F^2 = 4/27$ is a double point: one can easily verify that the response curve has a singular point with a double tangent at this point. The transition to response curves with no vertical tangents occurs for the response curve passing through the vertex $\sigma = \sqrt{3}/3, \rho = 2/3$ of the ellipse, and one finds for

---

* It may be of interest to contrast these occurrences with the analogous but considerably different occurrences in the case of forced oscillations of systems with nonlinear restoring forces as treated in Chapter IV. In the latter case we found that instabilities occurred only on the portions of the response curves which extended between two vertical tangencies (cf. Figure 3.1 in Chapter IV), and that no instability occurred on a response curve unless there was a vertical tangent on it. It follows that the jump phenomena encountered in the earlier case cannot exist in the same way in the problem now being considered. Nevertheless we shall see later that jump phenomena do occur in the present case; in particular jump phenomena involving a transition from an unstable harmonic oscillation to a stable nonperiodic oscillation, and vice versa, can and do occur in the present cases.

F^2 the value $F^2 = 8/27$. There is, finally, no difficulty in showing that the lines $\rho = \pm\sigma$ are tangent to the ellipse at $\sigma = \pm1/2, \rho = 1/2$ and that the response curve through the points $(\pm1/2, 1/2)$ is given by $F^2 = 1/4$.

In the next section we verify the correctness of the statements made above concerning the stability of the harmonic oscillations.

12. Stability of the harmonic oscillations

It has already been explained in detail in Section 10 that the stability of the harmonic oscillations, which has just been described in the preceding section in connection with the geometry of the response curves, is to be decided in accordance with the character of the singularities of the differential equation (10.4).* We turn then to the problem of classifying the singularities in accordance with the values ρ, σ, and F which determine a given harmonic oscillation. In order to do that we proceed as in Part B of Chapter III and replace x and y in (10.4) by $x_0 + \xi$, $y_0 + \eta$, with $x_0^2 + y_0^2 = \rho$ defined as the square of the amplitude of the harmonic oscillation whose stability is to be tested, develop the numerator and denominator in the right-hand side of (10.4) in powers of ξ and η, and reject all but the linear terms in ξ and η; the result is the following approximate differential equation in ξ and η:

$$(12.1) \qquad \frac{d\eta}{d\xi} = \frac{A\xi + B\eta}{C\xi + D\eta}.$$

This equation, according to a theorem of Poincaré, has the same types of singularities as the equation for x and y, provided that $AD - BC \neq 0$. In the present case one finds from (10.4) without difficulty that A, B, C, D have the following values:

$$(12.2) \qquad \begin{cases} A = \sigma - 2x_0 y_0, & B = 1 - \rho - 2y_0^2, \\ C = -2x_0^2 + 1 - \rho, & D = -\sigma - 2x_0 y_0. \end{cases}$$

* It would, of course, also be possible to investigate the stability of the harmonic oscillations by investigating the character of the solutions of the appropriate variational equation, in the manner outlined in Section 11 of Chapter IV for the Duffing equation. In the present case the resulting equation would be a Hill's equation with a periodic coefficient in the first derivative term.

In accordance with the discussion of Part B in Chapter III, we must calculate the quantities $(B - C)^2 + 4AD$, $AD - BC$, and $B + C$ in order to classify the singularities. We readily find for these quantities the expressions:

$$(12.3) \begin{cases} (B - C)^2 + 4AD = 4(\rho^2 - \sigma^2), \\ AD - BC = -(1 - \rho)(1 - 3\rho) - \sigma^2 = -E(\rho, \sigma), \\ B + C = 2(1 - 2\rho), \end{cases}$$

in which $E(\rho, \sigma)$ is the same quantity as given in (11.3); thus $E(\rho, \sigma) = 0$ is the equation of the ellipse indicated in the response diagram of Figure 11.1. It is now easy to determine the character of the singular points corresponding to all points of the response diagram—that is, corresponding to any given harmonic oscillation—simply by applying the criteria developed in Chapter III to the relations (12.3); we may summarize the result of such a classification as follows:

I. $\rho^2 > \sigma^2$ $\begin{cases} \text{(A) } E(\rho, \sigma) > 0 \begin{cases} \text{stable node} & \text{if } \rho > 1/2 \\ \text{unstable node} & \text{if } \rho < 1/2 \end{cases} \\ \text{(B) } E(\rho, \sigma) < 0, \text{ saddle} \end{cases}$

II. $\rho^2 = \sigma^2$ {node $\begin{cases} \text{stable} & \text{if } \rho > 1/2 \\ \text{unstable} & \text{if } \rho < 1/2 \end{cases}$

III. $\rho^2 < \sigma^2$ {center or spiral $\begin{cases} \text{stable} & \text{if } \rho > 1/2 \\ \text{unstable} & \text{if } \rho < 1/2 \end{cases}$

The decision regarding the stable or unstable character of a given singularity is based in Chapter III on the behavior of the solution curves as τ increases. The criteria given in the table therefore also hold for increasing values of the time t, since $\tau = t\alpha/2$ and $\alpha > 0$.

From this table we see that the saddle point singularities are confined to the interior of the ellipse $E(\rho, \sigma) = 0$, since $E < 0$ characterizes these points; the corresponding harmonic oscillations are therefore unstable. In the part of the ρ, σ-plane exterior to the ellipse it is clear from our table that the stable and unstable singularities are separated by the line $\rho = 1/2$, those below this line being unstable. This confirms the statements made in the preceding section regarding the stability of the various possible harmonic oscillations (cf. Figure 11.1, in which the unstable portions of the response curves are shown dotted). As we have already indicated, this result is due to van der Pol.

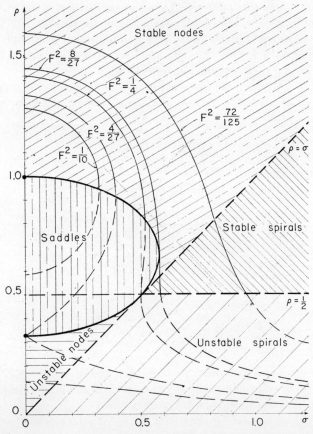

FIG. 12.1. Stability of the harmonic oscillations of a self-sustained system.

With reference to Figure 12.1, which superimposes the results of the above table classifying the singularities upon the response curves of Figure 11.1, we can summarize the situation with regard to harmonic oscillations of the van der Pol equation as follows:

1) If $F^2 > 8/27$ there is only one possible harmonic oscillation for a given value of σ (i.e. for a given frequency) and it is stable if $\rho > 1/2$,

i.e. if the square of its amplitude is larger than half the square of the amplitude of the free nonlinear oscillation (since $\sqrt{\rho} = b/a_0$), and is unstable for $\rho < 1/2$.

2) If $1/4 < F^2 < 8/27$ there is an interval of σ-values, i.e. $1/2 < \sigma < \sqrt{3}/3$, for which three different harmonic oscillations occur, of which either one or two will be stable depending upon whether $\rho > 1/2$ or not. The stable solution with larger amplitude corresponds to a nodal point. In Figure 12.2 we give a detail of a portion of Figure 12.1 indicating these circumstances.

3) If $F^2 < 1/4$, there are three harmonic oscillations for all values of σ between zero and a certain maximum (which depends on F),

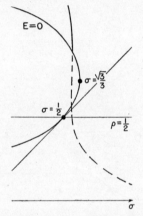

Fig. 12.2. Detail of Figure 12.1.

beyond which only one unstable harmonic oscillation occurs. When three periodic solutions occur only the one with the largest amplitude is stable, and it corresponds to a nodal point.

By making use of the character of the singularities one can also say something about the nature of the transient motions in the neighborhood of the stable harmonic oscillations: The harmonic oscillation corresponding to a given singularity can be represented (cf. (9.11)) in the form $v = b \cos (a_1 t + \varphi_0)$ by introducing an appropriate phase shift φ_0. If the singularity is a spiral point we could expect any motion in its neighborhood to be given approximately by $v = b \cos (\omega_1 t + \varphi_0) + \delta(t) \cos (\omega_1 t + \psi(t))$ in which the amplitude $\delta(t)$ of the disturbance tends to zero while the phase $\psi(t)$ tends to

$\pm \infty$ with increasing t. On the other hand if the singularity were a stable node, we would expect $\delta(t)$ to approach zero but $\psi(t)$ to tend to a finite limit as t increases. In other words, for $\rho < \sigma$ the transients would be oscillatory in character, but not for $\rho > \sigma$.

13. Nonharmonic response in general. Existence of stable combination oscillations for large detuning

Within the basic assumptions made and the accuracy of the approximation used here, the harmonic oscillations of the van der Pol equation have been completely characterized in the preceding sections, including their stability. In addition, it was possible to say something qualitative about the nonperiodic transients which may occur in the neighborhood of the stable harmonic oscillations. It is, however, possible to go much farther with the discussion of nonperiodic solutions of the type (9.1) of the van der Pol equation by considering the general character of the totality of solutions of the first order equations (9.8), which in turn fix the amplitude and phase of the former solutions. Such a discussion makes it possible to understand in some cases why, if at all, a given stable harmonic oscillation should be the motion observed in practice (after some transients have died out) rather than some other possibly nonperiodic motion. More than that, we see by reference to Figure 12.1 that there are no stable harmonic oscillations at all if $F^2 > 8/27$ and σ is sufficiently large compared with F: in other words if the detuning $\sigma = \Delta/\alpha$ is large enough no stable harmonic oscillation should be observed. This is actually what is found in practice. The important question then arises: what can be predicted from our basic theory for large detuning?

These and other questions can be answered in some instances, at least, by investigating the possibility of the occurrence of a *stable limit cycle* $\{x(\tau), y(\tau)\}$ (that is, a limit cycle toward which neighboring solutions tend as τ increases) among the solution curves of (10.3). As we have already indicated in Section 10, this would mean that the corresponding oscillation $v/a_0 = x \cos \omega_1 t + y \sin \omega_1 t$ would be affected by phase and amplitude modulation since x and y would be periodic functions of t with large periods, and, in addition, it would be stable. If such an oscillation were audible, beats would be heard. Such oscillations are thus somewhat like the combination tones studied in

Chapter IV and Appendix II since they contain two frequencies whose ratio is in general incommensurable; we shall therefore refer to them in what follows as *combination oscillations* to distinguish them from the harmonic oscillations. Such combination oscillations will in general be almost-periodic rather than periodic functions. The existence of combination oscillations in electrical circuits like that of Figure 8.1 is well known experimentally.

In the remainder of this section we shall discuss the technique of determining the existence of a limit cycle of (10.3) in a general way and show how it can be applied to prove that a stable limit cycle, and consequently a stable combination oscillation, occurs when the detuning is sufficiently large compared with the amplitude of the excitation. In the following Section 14 we shall then give an analytic treatment of the combination tones for large detuning whose existence is ensured because of the existence of a limit cycle. In Section 15 we consider the character of the oscillations for *sufficiently small detuning* and show, following Andronow and Witt [1], that *no limit cycles exist*, and hence no combination oscillations.

We turn then to the qualitative discussion of the solution curves of (10.3) in the entire x, y-plane, with particular attention to the possible occurrence of limit cycles. To begin with, the solution curves of (10.3) have an important property for large values of x and y which is independent of the values of the parameters σ and F (i.e. of the detuning and the amplitude of the excitation), i.e. that all solution curves of (10.3) for sufficiently large values of x and y are approximately straight lines through the origin, and x and y both tend toward the origin on these lines with increasing τ. The correctness of this statement follows at once from (10.3), since we have approximately for large x and y: $dx/d\tau = -r^2x$, $dy/d\tau = -r^2y$, so that $dy/dx = y/x$. Thus we see that the integral curves are approximately the rays through the origin and that a point $(x(\tau), y(\tau))$ on one of them moves toward the origin as τ increases. *Hence all integral curves of* (10.3) *remain, as τ increases, within a circle of sufficiently large radius.*

Within such a circle we know from the discussion of the preceding section that there are from one to three stable or unstable singularities of the differential equations (10.3), depending upon the values of the parameters σ and F, and these singularities include all of the possible types studied in Chapter III. One cannot, therefore, say anything concrete about the general character of the solution curves in the large since the variety of possibilities is much too great. However, in special cases it is quite feasible to decipher the general character of the solution curves, at least to the extent of proving the existence— or, under other circumstances, the nonexistence—of a limit cycle. For this purpose we shall have frequent occasion to make use of an important theorem of Poincaré and Bendixson, which furnishes conditions under which the existence of a limit cycle is ensured.

The theorem of Poincaré and Bendixson can be formulated as follows: If a solution curve $C: \{(x(t), y(t))\}$ of

$$dy/dx = P(x, y)/Q(x, y),$$

with P and Q defined and regular in $-\infty < x, y < \infty$, remains, as $t \to \infty$, within a bounded region of the x, y-plane without approaching singular points, then there exists at least one closed solution curve of the differential equation. For a proof of this theorem, see, for

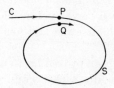

example, Bieberbach [5]. One can make the correctness of the theorem rather plausible in the following way. Consider an infinite sequence of points $P_i : (x(t_i), y(t_i))$ on C for $t_i \to \infty$. Such a set of points must have a limit point P since the points P_i lie in a bounded region, and P is not a singular point. If P lies on C it is not difficult to make it plausible geometrically that C itself is closed: One considers the accompanying figure and sees that the curve C must eventually cross the normal to it at P at a point Q so close to P that further passages of the curve across the normal as $t \to \infty$ would of necessity take place always in the same sense across the normal; but that is manifestly not possible since the solution curve beyond Q can escape out of the region $PSQP$ only by crossing over the segment PQ since an intersection with C itself is ruled out by our assumption that no singularities occur on C. Hence C itself is closed in this case. On the other hand, if P does not lie on C, one can make it very plausible by a similar geometrical argument that C winds itself in spiral fashion around a closed solution curve containing the point P.

The theorem of Poincaré and Bendixson can be applied, in particular, to establish the existence of a limit cycle when the detuning is large while the amplitude of the excitation is fixed, i.e. when σ is large compared with F.   As we see from Figure 12.1 there is only one singular point of (10.3) in the x, y-plane and it is an unstable spiral point when $\sigma >> F$ for any given F.   This means that any solution curve starting near this point moves away from it with increasing τ; in fact, there is an ellipse containing the spiral singular point in its interior with the property that all solution curves cross it on moving from its interior to its exterior as τ increases.   On the other hand we have seen that all solution curves which start on a circle of sufficiently large radius with center at the origin move into and stay inside the circle as τ increases.   Thus there is a ring-shaped domain bounded on the outside by this circle and on the inside by a small ellipse which is free of singular points and has the property that any solution curve which starts inside it remains inside it as $\tau \to \infty$.   The theorem of Poincaré and Bendixson can therefore be applied to establish the existence of *at least one* limit cycle and hence also of a combination oscillation.

It seems not to have been proved up to now that there is *only one* such cycle if σ is sufficiently large; we shall give such a proof in Section 16.   Once the uniqueness of the limit cycle has been proved, it is clear that the limit cycle is stable,* and hence that the combination oscillation is also stable.   We see, therefore, that the approach of Andronow and Witt leads to a proof of the existence of a unique stable combination oscillation for large detuning.   In the next section we discuss such oscillations quantitatively by an analytic approach which was already used by van der Pol.

14. *Quantitative treatment of combination oscillations for large detuning*

Our interest is in cases in which the detuning σ is made large while the amplitude of the excitation is held fixed.

In this section there is some advantage in returning for a time to the original notations of Sections 8 and 9.   In terms of the quantities

---

* Curiously enough, the uniqueness of the limit cycle is established in Section 16 by proving first that all limit cycles which might occur would be stable, after which the uniqueness of the cycle follows with no difficulty.

used there our assumption may be interpreted to mean that Δ is large compared with B in equations (9.8). Let us begin our discussion by assuming that Δ is actually so large that the amplitude $b = \sqrt{b_1^2 + b_2^2}$ of the resulting oscillation is small compared with it. In such cases equations (9.8) may clearly be written in the following approximate form:

$$\begin{cases} 2\dot{b}_1 + b_2\Delta = 0, \\ 2\dot{b}_2 - b_1\Delta = 0, \end{cases}$$

which have as solutions $b_1(t) = b_f \cos{(\frac{\Delta}{2} t + \varphi_0)}, b_2(t) = b_f \sin{(\frac{\Delta}{2} t + \varphi_0)}$

The solution (9.1) of equation (8.4) therefore can be written in this case in the form

$$(14.1) \quad v(t) = b_f \sin\left(\omega_1 t + \frac{\Delta}{2} t + \varphi_0\right) = b_f \sin{(\omega_0 t + \varphi_0)},$$

the last step following from the definition of Δ: $\Delta = 2(\omega_0 - \omega_1)$ given by (9.7). The quantity φ_0 is a constant phase shift. The quantity b_f, the amplitude of the oscillation, is written with the subscript f because *the oscillation evidently has the frequency ω_0 of the free oscillation* of the system. In the limit, as the detuning grows large, we shall see later that b_f also actually tends to the amplitude a_0 of the free oscillation.

In other words, we expect the oscillation of the system to be approximated closely by the free oscillation when the detuning is very large. On the other hand, if the detuning is very small we observe from Figure 12.1 that there is always one and only one stable harmonic oscillation for any given value of the excitation amplitude, and we shall show in the next section that it is the only stable motion (aside from transients) which exists for small detuning. At the two extremes of large and small detuning the resulting oscillations are thus simple harmonic oscillations which have the frequencies ω_0 and ω_1 respectively. Between the two extremes we know from the preceding section that there is a range of values of the detuning for any given amplitude of the excitation within which combination oscillations of the form $v(t) = b_1 \sin \omega_1 t + b_2 \cos \omega_1 t$ occur with $b_1(t)$ and $b_2(t)$ certain periodic functions having a common frequency. It would seem at least plausible that the oscillations for moderately large values of σ—more precisely those values of σ for which the harmonic oscillation corresponds (cf. Figure 12.1) to an unstable spiral point in the b_1, b_2-phase plane—could be considered as approximately a combination of two simple harmonic oscillations, one with the frequency ω_0 of the free oscillation, the other with the frequency ω_1 of the excitation. In order to investigate the combination oscillations quantitatively it would therefore seem reasonable to replace the general form $v = b_1 \sin \omega_1 t + b_2 \cos \omega_1 t$ of the combination oscillation by a sum of two simple harmonic oscillations, as follows:

$$(14.2) \qquad v(t) = b_f \sin (\omega t + \varphi_1) + b_h \sin (\omega_1 t + \varphi_2).$$

The quantities b_f, b_h, φ_1, φ_2, and ω are all constants with a fairly obvious significance: b_f and b_h are the amplitudes of the "free" and the "harmonic" components of the combination oscillation, respectively, while ω and ω_1 are their frequencies. We have preferred not to set ω equal to ω_0, the frequency of the free oscillation, at the outset since it is conceivable that the frequency of this component might turn out to differ somewhat from ω_0—actually we shall see that $\omega = \omega_0$ within the accuracy of our approximation. The constants in (14.2) must now be chosen in such a way that (14.2) yields $v(t)$ as a solution of the original differential equation (8.4); by substituting $v(t)$ in (8.4) and making the usual approximations one obtains the desired relations for the constants by equating the coefficients of the terms in $\sin \omega t$, $\cos \omega t$, $\sin \omega_1 t$, and $\cos \omega_1 t$. The result is the following set of equations:

$$(14.3) \begin{cases} \text{a)} \quad (\omega_0^2 - \omega^2)b_f = 0 \\[2mm] \text{b)} \quad \omega \alpha b_f \left(1 - \dfrac{b_f^2 + 2b_h^2}{a_0^2} \right) = 0 \\[2mm] \text{c)} \quad b_h(\omega_0^2 - \omega_1^2) \cos \varphi_2 + \alpha b_h \omega_1 \left(1 - \dfrac{b_h^2 + 2b_f^2}{a_0^2} \right) \sin \varphi_2 = B\omega_0^2 \\[2mm] \text{d)} \quad b_h(\omega_0^2 - \omega_1^2) \sin \varphi_2 - \alpha b_h \omega_1 \left(1 - \dfrac{b_h^2 + 2b_f^2}{a_0^2} \right) \cos \varphi_2 = 0, \end{cases}$$

in which a_0 is, as before (cf. (9.6)), the amplitude of the free nonlinear oscillation. From the first of the relations (14.3) we conclude that if $b_f \neq 0$, then $\omega = \omega_0$; in other words the first term on the right-hand side of (14.2) is an oscillation with the frequency of the free nonlinear oscillation, as we had expected.

We proceed next to a study of the amplitudes b_f and b_h of the components of the combination oscillation.   If $b_f = 0$, it is at once seen that the relations (14.3) reduce, as they should, to the corresponding relations for the harmonic oscillations studied in Section 11, and b_h can be identified with the quantity b used to denote the amplitude of the harmonic oscillation.   From (14.3b) we find, in case $b_f \neq 0$:

$$(14.4) \qquad\qquad b_f^2 + 2b_h^2 = a_0^2,$$

while (14.3c) and (14.3d) yield

$$(14.5) \qquad\qquad \frac{b_h\,\sigma}{a_0} = F \cos\varphi_2$$

and

$$(14.6) \qquad\qquad \frac{b_h}{a_0}\left(1 - \frac{b_h^2 + 2b_f^2}{a_0^2}\right) = F \sin\varphi_2$$

with F and σ defined as in (10.2).

From now on we work in general once more with the quantities introduced in Section 10.   As a measure of the square of the "amplitude" of the oscillation given by (14.2) we take the quantity ρ defined by

$$(14.7) \qquad\qquad \rho = \left(\frac{b_f}{a_0}\right)^2 + \left(\frac{b_h}{a_0}\right)^2 = \left(\frac{b_f}{a_0}\right)^2 + \rho'$$

so that ρ' is defined by

$$(14.8) \qquad\qquad \rho' = \left(\frac{b_h}{a_0}\right)^2$$

and is the square of the amplitude of the harmonic component. We have already seen that $b_h = b$ if $b_f = 0$; in this case therefore the quantity ρ as defined here is identical with the same quantity used in Section 11 to discuss the harmonic oscillations.   From (14.4) we have

$$(14.9) \qquad\qquad \left(\frac{b_f}{a_0}\right)^2 = 1 - 2\rho'$$

so that ρ' can never exceed the value $1/2$, at which $b_f = 0$ and hence also $\rho = \rho'$, $b = b_h$.   By comparing with Figure 12.1 we note that the transition from stable to unstable harmonic oscillations occurs exactly for $\rho = 1/2$, at which $b_f = 0$.

We now imagine an experiment performed in which F (proportional to the amplitude of the excitation) is held fixed while σ, the detuning, is slowly increased. In view of our earlier discussions we expect that an oscillation of the type given by (14.2) with two harmonic components sets in as soon as σ is increased beyond the point on the harmonic response curve where $\rho = 1/2$ (cf. Figure 12.1). For such values of σ we expect a nonharmonic response with an "amplitude" ρ given by (14.7). It is convenient to superimpose the response curves for the harmonic oscillations on those for the combination oscillations which result from (14.4), (14.5), and (14.6), the latter being valid only for the range in which the combination oscillations exist, which means in the region $\rho < 1/2$ where the harmonic oscillation is not stable. The quantity ρ is thus unambiguous in the sense that it furnishes always the amplitude of whatever stable oscillation exists. The "amplitude" of the combination oscillation as a function of the detuning σ for any given F can be discussed by means of the equation

$$(14.10) \qquad\qquad \rho = 1 - \rho',$$

which follows from (14.7) and (14.9), and the equation

$$(14.11) \qquad\qquad \rho'[\sigma^2 + (1 - 3\rho')^2] = F^2,$$

which follows from (14.5) and (14.6) by elimination of φ_2 and $(b_f/a_0)^2$. The relation (14.11) yields through the value of ρ' the amplitude of the "harmonic" fraction of the response. By comparison with (11.2) we see that this portion of the response is given by the same expression as in the case of the pure harmonic oscillation except for the factor 3 on ρ' inside the bracket. In any case, it is clear from (14.11) that $\rho' \rightarrow 0$ as $\sigma \rightarrow \infty$ for any fixed value of F, so that $\rho \rightarrow 1$ from the lower side as $\sigma \rightarrow \infty$, which confirms the observation at the beginning of this section that the motion to be expected for very large σ is the free oscillation. The response curves obtained from (14.10) and (14.11) are shown in Figure 14.1, together with the stable portions of the harmonic response curves. The branches of the response curves which rise from $\rho = 1/2$ as σ increases correspond to the response in the form of combination oscillations, while the branches which go downward to $\rho = 1/2$ as σ increases correspond, of course, to the harmonic oscillations discussed in Section 11.

The present section was concerned with the analysis of the limit case in which σ is large, i.e. with the case of large difference between the frequency of the free oscillation and the excitation for a given value of the amplitude of the excitation.   From Figure 14.1 we see also that F^2, which characterizes the amplitude of the excitation, should be larger than 8/27 in order to obtain response curves of the kind shown in Figure 14.1.   In the next section we shall analyze the opposite limit case, in which σ and F are both small, and finally in the

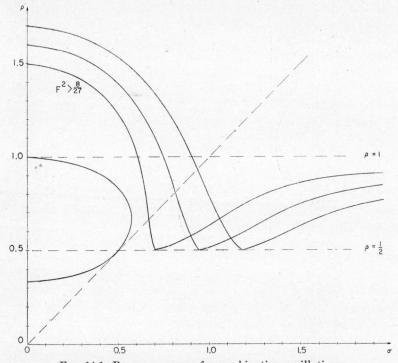

Fig. 14.1. Response curves for combination oscillations.

next to the last section of this chapter we shall give a brief description of the more complicated circumstances encountered in the middle region.

15. Nonexistence of combination oscillations when the detuning and the amplitude of the excitation are sufficiently small

The situation is now very different from that of the preceding section since there are now three different harmonic oscillations possible: one corresponding to a stable nodal point, one corresponding to a saddle point, and one corresponding to an unstable nodal point in the x, y-plane, as we have seen in Section 12. This is indicated in Figure 15.1, which is a detail of Figure 12.2. We note that the stable nodal point occurs for $\rho > 1$, the saddle point for $1 > \rho > 1/3$, and the unstable nodal point for $\rho < 1/3$ if σ is small enough. The

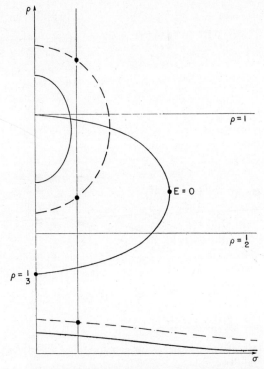

Fɪɢ. 15.1. Response curves for small detuning.

question now is: what sort of stable motions does our theory predict for this range of parameters? When we consider the results of the preceding section, we are led to inquire whether it might not be possible that a limit cycle of the differential equation (10.4) exists in the present cases also, at least for certain values of the parameters σ and F, in which case combination tones would exist. This is, however, not the case, and *stable motions exhibiting perfect synchronization*

with the frequency of the excitation occur no matter how small the ampli-
tude F of the excitation may be, provided only that the detuning σ is
kept under a certain bound which depends on F. This bound $\bar{\sigma}$ should
be fixed so that the stable nodal points for all $\sigma < \bar{\sigma}$ occur for $\rho > 1$,
as indicated in Figure 15.1. This is an experimentally verified fact
in electrical circuits of the type discussed in Section 8. In the
present section we shall see that these observations are in accord
with our theory by using an argument given by Andronow and Witt
[1] in the paper cited above; in fact, Andronow and Witt had as the
main object of their investigation the giving of a proof that there is no
threshold value for the amplitude of the excitation below which
synchronization does not occur.

In view of the discussion above, particularly that of the preceding
section, we see that what is necessary to achieve the desired result
in the present case is a proof that *no limit cycle of* (10.4) *can occur in*
the x, y-plane when σ is kept under the bound $\bar{\sigma}(F)$ discussed above:
If this is once established it is clear that the only stable motion
(excluding transients) which occurs is the harmonic oscillation
corresponding to the stable nodal point of (10.4), in view of the
theorem of Poincaré and Bendixson (see Section 13 for the formula-
tion of this theorem), according to which every solution of equation
(10.4) would of necessity approach the stable singular point: We
have already seen in Section 13 that all solution curves remain inside a
sufficiently large circle when the curve parameter τ (essentially the
time) becomes and remains sufficiently large. Hence all solution
curves must tend either to a limit cycle or to a stable singular point,
and since there is only one stable singular point in the present case all
solution curves must tend to it as the time increases.

We have then to show that no limit cycles occur. The general
method used by Andronow and Witt to prove this is the following:
One first constructs two ring domains, each centered at one of the
two different nodal points of the x, y-plane, with the property that
any limit cycles which occur lie in the interiors of the ring domains
and hence in their common part. In other words, a construction is
used which traps the limit cycles in certain regions R of the plane.
Each region R is then shown to be simply connected, i.e. to be such
that every closed curve in it can be shrunk to a point without leaving
the region. Once these constructions have been made, it is readily
seen that no limit cycle occurs at all, as follows. If there were such a

cycle it would contain in its interior either no singular point or a saddle point, since the remaining singular points have been excluded by the above construction from the simply connected region R which contains the cycle. On the other hand, a closed solution curve must contain at least one singular point of index >0 in its interior, as we have seen in Section 7 of Chapter III. The assumption that there is a limit cycle in R must therefore be rejected since such a cycle would contain either no singularity or a singularity of index -1.

The object of this section will therefore be accomplished once the two ring domains with the properties described above have been constructed. The method used by Andronow and Witt for this purpose involves the determination of certain curves introduced by Poincaré and called *contact curves* by him. For the present purposes they are defined as follows: We consider a set of concentric circles with centers at a singular point (x_0, y_0) of (10.4) with radii r_1 and determine the locus of the points where these circles are tangent to the field directions determined by (10.4); this locus is the contact curve. The ring region centered at (x_0, y_0) in which any possible limit cycles occur is then shown to be determined by the circles of largest and smallest radius which touch the contact locus. The relative positions and sizes of the ring domains centered at the two nodal singularities must then be studied in detail in order to prove that the region common to both of them is simply connected. We turn then to the task of carrying out the details of this program.

It is convenient to begin by assuming that the contact loci lie in a bounded region of the plane—a fact that will be apparent later on—and show on the basis of this assumption that *if a limit cycle occurs at all it must of necessity lie in a ring with center at the singular point* (x_0, y_0) *whose boundaries are the innermost and outermost circles of radii* r_{1min} *and* r_{1max} *respectively, which touch the contact curve.* The proof is carried out in three steps. First of all we show that there exist points of the contact locus in every neighborhood of the two other singular points. We observe that one of these, the node, has the index $+1$, while the other (the saddle point) has the index -1, which means that the direction field in the neighborhood of these singularities turns through the angle $+2\pi$ or -2π (cf. Chapter III, Section 7) on making a circuit around every sufficiently small circle with center at the singularity. It is therefore clear that the circles with centers at (x_0, y_0) would have at least one common direction with the field in every neighborhood of some other singularity since the field of directions defined by the concentric circles is essentially a parallel field in the neighborhood of each of the singular points (or, as one could also put it, the index of this field is 0 everywhere except at the point (x_0, y_0)). The second step in proving our statement is to show that the point (x_0, y_0) at which the ring domain is centered is *an isolated point of the contact locus.* This is seen as follows: At a nodal point $(0, 0)$ the direction field is given essentially by $dy/dx = a(y/x)$, $a > 0$ or by $dy/dx = (y + x)/x$, as we have seen in Chapter III, while the circles centered at $(0, 0)$ have the slopes $dy/dx = -x/y$; one sees therefore that the only point where the respective right-hand sides are equal is the origin since $ay^2 + x^2$ and $x^2 + xy + y^2$ as positive definite forms vanish only for $x = y = 0$. One sees also that the argument would still hold if the higher order terms in (x, y) in the differential equations for the field directions were retained. The third step in proving our statement is then the following: If there were a point of a limit cycle *outside* the circle of radius r_{1max} say, it follows that there would be a point P_{max} on the limit cycle at a maximum distance R_{max} from (x_0, y_0), with $R_{max} > r_{1max}$. If P_{max} were not a singular point, the limit cycle would have a tangent at P_{max} which would also clearly be a tangent to the circle of radius R_{max} centered at (x_0, y_0). If P_{max} should happen to be a singular point, we know from the discussion above that there would be tangencies of the circles centered at (x_0, y_0) with the integral curves in any arbitrary neighborhood of P_{max}. In either case, therefore, there would be points of the contact locus outside the circle of radius r_{1max}, contrary to our assumption. In the same way we can show that if a point of a limit cycle were to lie inside the circle of radius r_{1min} it would follow that points of the contact locus would occur near a point P_{min} of the limit cycle at the minimum distance from (x_0, y_0), with $P_{min} \neq (x_0, y_0)$ since (x_0, y_0) was shown above to be an isolated point of the contact locus. In other words, the assumption that there are points of a limit cycle outside the region in question leads to a contradiction, and hence our statement is proved. We have therefore a means of trapping the possible limit cycles in definite regions of the plane.

To make a study of the contact curves as a means of locating possible limit cycles of (10.4) it is convenient to shift the origin to a singularity (x_0, y_0) by introducing $x_1 = x - x_0$, $y_1 = y - y_0$ as new variables; the result is

$$(15.1) \quad \frac{dy_1}{dx_1} = \frac{F + \sigma(x_1 + x_0) + (y_1 x + y_0)(1 - r^2)}{-\sigma(y_1 + y_0) + (x_1 + x_0)(1 - r^2)} = \frac{P(x_1, y_1)}{Q(x_1, y_1)},$$

with $r^2 = (x_1 + x_0)^2 + (y_1 + y_0)^2$. The contact curves can be conveniently obtained by transforming (15.1) to polar coordinates (r_1, ψ) with center at $(x_1, y_1) = (0, 0)$; the appropriate formula is

$$(15.2) \quad \frac{1}{r_1} \frac{dr_1}{d\psi} = \frac{x_1 Q + y_1 P}{x_1 P - y_1 Q}, \qquad \text{with } r_1^2 = x_1^2 + y_1^2.$$

The contact curves are evidently given by $dr_1/d\psi = 0$, so that they are furnished by the algebraic curves $x_1 Q + y_1 P = 0$ or

$$(15.3) \quad \begin{aligned} x_1[-\sigma(y_1 + y_0) + (x_1 + x_0)(1 - r^2)] \\ + y_1[F + \sigma(x_1 + x_0) + (y_1 + y_0)(1 - r^2)] = 0, \end{aligned}$$

together with $r_1 = 0$.

The fact that the origin (in the x_1, y_1-plane) is a singular point imposes the conditions $P(0, 0) = Q(0, 0) = 0$ on σ, F, x_0, and y_0. These conditions are of course the same as those given in Section 11:

$$(15.4) \quad \begin{cases} -\sigma y_0 + x_0(1 - \rho) = 0, \\ F + \sigma x_0 + y_0(1 - \rho) = 0 \end{cases} \qquad \rho = x_0^2 + y_0^2.$$

From these we have already deduced in Section 11 the relation

$$(15.5) \quad \rho[\sigma^2 + (1 - \rho)^2] = F^2,$$

from which the response curves of Figure 11.1 were derived. It is useful for later purposes to give the following two further relations which are easily deduced from (15.4):

$$(15.6) \quad \begin{cases} x_0 = \dfrac{-\rho\sigma}{F}, \\ y_0 = -\dfrac{\rho(1 - \rho)}{F}. \end{cases}$$

The equations (15.6) determine the coordinates of the singularity in the x, y-plane, and of course also the center of the ring domain which we propose to construct.

By using (15.4) it is easily found that equation (15.3) reduces to

(15.7) $r_1^2(1 - r^2) - (x_1x_0 + y_1y_0)(r_1^2 + 2x_1x_0 + 2y_1y_0) = 0,$

upon eliminating F and σ. We introduce the quantity κ by the relations

(15.8) $\begin{cases} x_1x_0 + y_1y_0 = r_1(x_0 \cos \psi + y_0 \sin \psi) = r_1\kappa \\ \kappa = x_0 \cos \psi + y_0 \sin \psi \end{cases}$

and note that $r^2 = r_1^2 + \rho + 2r_1\kappa$, so that (15.7) can finally be written in the form:

(15.9) $r_1^2[r_1^2 + 3\kappa r_1 - (1 - \rho - 2\kappa^2)] = 0.$

Equation (15.9) is the equation of the contact curve in polar coordinates. Aside from the point $r_1 = 0$, the contact curve is given by

(15.10) $r_1 = \dfrac{-3\kappa}{2} \pm \sqrt{\dfrac{\kappa^2}{4} - \rho + 1};$

and we observe that *it has no branch which goes to infinity.* Consequently we know that a ring domain centered at the singular point (x_0, y_0) exists and contains all possible limit cycles; its boundaries are the radii $r_{1\min} > 0$ and $r_{1\max}$ of the smallest and largest circles which touch the contact locus given by (15.10). The radii of these circles are contained among the roots of the equation $dr_1/d\psi = 0$. It follows easily that these values of r_1 satisfy one or the other of the relations

(15.11) $\begin{cases} 3r_1 + 4\kappa = 0, \quad \text{or} \\ \dfrac{d\kappa}{d\psi} = 0. \end{cases}$

The first of these relations, upon insertion in (15.9) or (15.10), yields for r_1 the value

(15.12) $r_1 = \sqrt{8(\rho - 1)}.$

The second relation yields by (15.8) the equation

$$-x_0 \sin \psi + y_0 \cos \psi = 0$$

and hence $\kappa = \sqrt{\rho}$ since $\rho = x_0^2 + y_0^2$. Insertion of $\kappa = \sqrt{\rho}$ in (15.10) leads then to the value

(15.13) $r_1 = \left| \dfrac{3}{2} \sqrt{\rho} \pm \sqrt{1 - \dfrac{3}{4}\rho} \right|$

for r_1. According to our discussion above, any existing limit cycle lies in the ring between the circles of radii $r_{1\text{max}}$ and $r_{1\text{min}}$ obtained by taking the largest and smallest values furnished by (15.12) and (15.13).

We now apply the above considerations to the special case of interest to us here in which both F and σ are below certain bounds. We must discuss the ring regions centered at the two nodal points which occur in the present case by studying the values of $r_{1\text{max}}$ and $r_{1\text{min}}$ as well as the location of the centers of the rings as functions of the two parameters F and σ. In order to distinguish the two different rings and the quantities associated with them we use the subscripts s and u, the former referring to the ring associated with the stable nodal point and the latter to the unstable nodal point. Thus $r_{0u} = \sqrt{x_{0u}^2 + y_{0u}^2} = \sqrt{\rho_u}$ and $r_{0s} = \sqrt{\rho_s}$ represent the distances in the x, y-plane to the corresponding singularities, while $r_{1u\ \text{max}}$ and $r_{1s\ \text{max}}$ represent the radii to the outer boundary circles of the rings centered at the corresponding singularities. Finally (x_{0u}, y_{0u}) and (x_{0s}, y_{0s}) represent the centers of the rings—in other words the singular points themselves. We now determine all of these quantities approximately for small F and σ by using equations (15.5) and (15.6) to determine the centers of the rings, after which their radii can be determined from (15.12) and (15.13). To this end consider first the unstable nodal point. From Figure 15.1 we know that ρ_u is less than $1/3$ at this point if F^2 is less than $4/27$; it then follows from (15.5) that ρ_u and r_{0u} have the approximate values

$$(15.14) \qquad \rho_u = F^2, \qquad r_{0u} = F$$

within higher order terms in F if σ is small enough, and this approximation is better the smaller F is taken. As for the coordinates (x_{0u}, y_{0u}) of the center of the ring, one sees from (15.6) and (15.14) that they are given approximately by

$$(15.15) \qquad \begin{cases} x_{0u} = -F\sigma, \\ y_{0u} = -F. \end{cases}$$

At the stable nodal point we find in the same manner that ρ_s and r_{0s}, which are slightly greater than unity, are given approximately for small values of σ by

$$(15.16) \qquad \rho_s = 1 + F - \tfrac{1}{2}F^2 \qquad r_{0s} = 1 + \frac{F}{2}$$

again within a higher order term in F. The coordinates (x_{0s}, y_{0s}) are then given approximately by

(15.17)
$$\begin{cases} x_{0s} = -\dfrac{(1 + F)}{F}\, \sigma, \\[2mm] y_{0s} = 1 + \tfrac{1}{2}F. \end{cases}$$

We observe that the x-coordinates (but not the y-coordinates) of the centers of the two rings are of first order in σ.

We proceed to study the ring domain centered at the unstable nodal point. Since $\rho_u < 1/3$, it follows that the maxima and minima of r_{1u} are both obtained from (15.13) since (15.12) does not yield a real value for r_{1u}. The quantities $r_{1u\,max}$ and $r_{1u\,min}$ are thus obtained in terms of F by inserting the value $\rho_u = F^2$ from (15.14) for the quantity ρ in (15.13). These values are then easily seen to be given approximately by

(15.18)
$$\begin{cases} r_{1u\,max} = 1 + \tfrac{3}{2}\,F, \\[2mm] r_{1u\,min} = 1 - \tfrac{3}{2}\,F \end{cases}$$

for F small. Again second order terms in F have been ignored. Since the coordinates of the center are given by (15.15) we see that the ring domain associated with the unstable singular point appears as in Figure 15.2a, in which, however, the center is taken on the y-axis instead of slightly to the left of it.

In the same manner we determine the values of $r_{1s\,max}$ and $r_{1s\,min}$ approximately in terms of F. In this case one sees readily from (15.16) that $r_{1s\,max}$ is given by (15.13) and has the approximate value $r_{1s\,max} = 2$, but that $r_{1s\,min}$ might be furnished by either of the two expressions (15.12) or (15.13). An easy calculation yields the approximate values

(15.19)
$$\begin{cases} r_{1s\,max} = 2, \\[2mm] r_{1s\,min} = 1 + \tfrac{3}{2}\,F, \qquad \text{or} \qquad r_{1s\,min} = \sqrt{8F}, \end{cases}$$

once more within terms of higher order in F. In Figure 15.2b we indicate the circumstances in this case by showing the ring centered at distance $r_{0s} = 1 + F/2$ from the origin and with its center on the y-axis (cf. (15.16) and (15.17)). Only the inner radius of the ring for the case $r_{1s\,min} = \sqrt{8F}$ is shown, since the outer radius is so large.

Finally, we have only to superimpose the two ring domains and study the region which they have in common. In Figure 15.3 the two domains are shown together for both of the cases which may occur on account of the ambiguity in the values for $r_{1s\ min}$. The circle of radius $r_{0s\ max} = 2$, the outer boundary of one of the rings, is drawn only in part. The region common to the two ring domains is shaded in Figure 15.3. We can now see from Figure 15.3 that the domains common to the two ring domains are indeed simply connected

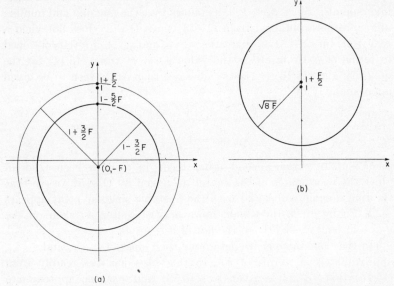

Fig. 15.2. Ring domains associated with each of the nodal singular points.

because of the fact that the center of the ring associated with the stable nodal point lies practically on the outer boundary of the other ring, while its smaller radius is larger than the distance, $3F$, between the circles centered at the unstable node. This is evident in the case shown in Figure 15.3b. In the case of Figure 15.3a the radius of the circle corresponding to the stable nodal point is $\sqrt{8F}$, which can be made large compared with F if F is small enough. The shaded regions shown in Fig. 15.3 may form one or two simply connected regions depending on slight changes in F or σ. For F sufficiently small*

* Actually, it could be seen that it is sufficient to choose F so that $F^2 < 4/27$.

we may thus conclude that Figure 15.3 is qualitatively correct, if it is legitimate to take the centers of the ring domains on the y-axis instead of slightly to the left of it in accordance with (15.15) and (15.17). It is, however, clear that the qualitative result would not be changed if we assume σ to be different from zero, but small: the effect upon Figure 15.3 would be simply to shift the centers of the domains to the left of the y-axis by amounts which are of first order in σ. If therefore F is chosen small enough so that Figure 15.3 is qualitatively correct for $\sigma = 0$, it follows that it will remain so for all

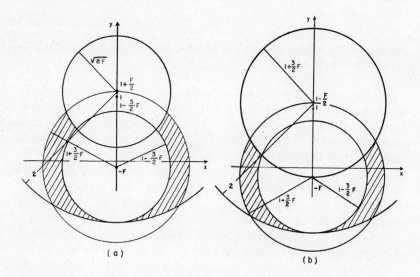

(a)

(b)

FIG. 15.3. Simply connected domain common to the two ring domains.

values of σ below a certain bound. We conclude therefore that *once F has been fixed at an appropriate small value the domains in which any limit cycles lie are simply connected if σ is below a certain bound $\bar{\sigma}(F)$.* This result, finally, ensures that no limit cycles exist for $\sigma < \bar{\sigma}(F)$, as we have already shown above.

No attempt was made in the above discussion to fix a *maximum* for the value of F at which a bound for σ exists such that no limit cycles of (10.4) exist. However, it could be shown that this result holds for all F. Andronow and Witt show that the discussion above (valid only for F sufficiently small) can be extended for values of F

such that $F^2 \leq 4/27$ (as one might expect). If $F^2 > 4/27$ so that only one singular point—a stable nodal point—occurs for σ sufficiently small (cf. Figure 12.1) we may apply the same kind of argument as above but confined now to the ring centered at the one singular point. For $F^2 > 4/27$ it is readily seen from (15.5) that ρ_s^2 is greater than $4/3$ if σ is small enough and hence that the contact curve furnished by (15.10) has no real branches, since $\kappa^2 < \rho_s$, as one sees from (15.8) and $\rho_s = x_{0s}^2 + y_{0s}^2$. Consequently no limit cycles exist if σ is sufficiently small. In other words there is a range of values of σ near $\sigma = 0$ for any F such that the harmonic oscillation is the only stable oscillation.

16. Stability and uniqueness of the combination oscillations for large detuning

In the preceding section it was shown that no combination oscillations of our system occur if the detuning σ is kept sufficiently small, by showing that no limit cycles of (10.3) occur; in the present section we employ some of the ideas developed in that section as an aid in showing that there is *a unique stable combination oscillation if the detuning is sufficiently large* compared with the amplitude of the excitation, by showing that only one stable limit cycle of (10.3) occurs. We have seen previously in Section 13 that at least one combination oscillation exists.

The proof of our statement takes the following course. We narrow down the location of all possible limit cycles of (10.3) by making use of the contact curves in the manner of the preceding section. In particular, we shall show that all limit cycles lie outside a circle in the x, y-plane whose center is very close to the origin and whose radius is nearly unity. We show next that any such cycle is stable and isolated by using Poincaré's criterion for stability, which is developed in Appendix V. The application of the criterion proves to be very simple in the present case. Finally we make use of an idea of Levinson and Smith [24] to conclude from the fact that all possible cycles are isolated and stable, that there is only one cycle. In Appendix VI this conclusion is established, and we shall not repeat the proof here. In other words, we have only to show that any possible cycles are stable and isolated in order to be sure that only one cycle exists, and hence that there is only one combination oscillation, which is then also evidently stable.

We turn to the proof that any limit cycles of the equations (10.3) are isolated and stable if σ is large compared with F. It is convenient to write down these equations once more:

$$(16.1) \quad \begin{cases} \dfrac{dx}{d\tau} = -\sigma y + x(1 - x^2 - y^2) = P(x, y), \\[2mm] \dfrac{dy}{d\tau} = F + \sigma x + y(1 - x^2 - y^2) = Q(x, y). \end{cases}$$

In the present case, in which $\sigma \gg F$, there is only one singular point of (16.1)—an unstable node—and it is located very near to the origin in the x, y-plane. The coordinates (x_0, y_0) of the singular point and the quantity $\rho = x_0^2 + y_0^2$ satisfy the relations (15.6) and (15.5) of the preceding section. If we set $\epsilon = F/\sigma$ we find without difficulty for these quantities the values

$$(16.2) \qquad \rho = \epsilon^2, \qquad x_0 = -\epsilon, \qquad y_0 = -\frac{\epsilon^2}{F}; \qquad \epsilon = F/\sigma,$$

within quantities of higher order in the small quantity ϵ. We consider next the ring domain centered at the singular point (x_0, y_0) which contains any possible limit cycles of (16.1). The inner and outer radii $r_{1\max}$ and $r_{1\min}$ of the circles bounding the ring are furnished, just as in the preceding section, by the largest and smallest values of r_1 given by equation (15.12) or equation (15.13). In the present case, in which ρ is small, the equation (15.12) does not give a real value for r_1; consequently both radii are furnished by (15.13). With the value $\rho = \epsilon^2$ for ρ we then see at once that $r_{1\min} = 1$ *within first order terms in* ϵ. From (16.2) we observe also that the center of the ring domain is at a distance from the origin which is of first order in ϵ. Hence if σ is sufficiently large compared with F—and hence ϵ sufficiently small—it is clear that any existing limit cycles lie outside of a circle with center nearly at the origin and of radius nearly unity. For establishing the stability criterion it will be sufficient to know that *all limit cycles lie outside a circle of radius* 3/4 *with center at the origin.* Clearly this is true if σ is sufficiently large.

We have now to show that all cycles satisfy the stability criterion given in Appendix V. The criterion for an isolated and stable limit cycle is the following inequality:

$$(16.3) \qquad \oint (P_x + Q_y)d\tau < 0.$$

The integral is taken in the direction of increasing τ over the cycle whose stability is to be tested, and P_x and Q_y are the partial derivatives of the functions which figure in the right-hand sides of the equations (16.1). In the present case we have

$$(16.4) \qquad \oint (P_x + Q_y)d\tau = \oint [2 - 4(x^2 + y^2)]d\tau,$$

and the integral will obviously be negative if $(x^2 + y^2) > 1/2$ over the entire cycle. But we know from our discussion above that any existing cycles lie outside the circle $x^2 + y^2 = (3/4)^2$, and hence the criterion for stability is satisfied. Once this is known, the uniqueness of the cycle is also established, as was remarked above.

We have therefore proved that there is one and only one combination oscillation, which is in addition stable, provided that the detuning is sufficiently large compared with the excitation amplitude. It would be possible to refine this result in such a way as to give more precise bounds for the parameter values within which unique combination oscillations occur.

17. Description of the response phenomena for intermediate values of the detuning σ. Jump phenomena.

In Section 13 and in the preceding section we studied the nature of the response to be expected when the detuning σ is large compared with F, and in Section 15 the same question for sufficiently small F when σ is small compared with F. In the first limit case we found the stable oscillation to be composed of a sum of two oscillations, one with the frequency of the excitation, the other with a frequency approximately the same as that of the free nonlinear oscillation; in the second limit case we found the stable oscillation to be an oscillation with the frequency of the excitation. In the present section we give a brief description of the results for the intermediate cases in which the detuning σ is neither large nor small compared with F, following the work of Cartwright and Littlewood as reported, without proofs, in a recent paper by Cartwright [6]. These results are stated to hold rigorously within terms of the order retained in the theory as developed in Section 9 above.

In Figure 17.1 we reproduce the figure given by Cartwright (and which is the same as Figure 11.1, *in so far as the harmonic oscillations are concerned*) for the response curves in the portion of the response plane of interest here. The full curves correspond to the stable oscillations. As one sees, the response curves for $F^2 > 8/27$ and σ large are like those of Figure 14.1 and are to be interpreted in the same way: the rising branches of the response curves (with increase of σ) represent oscillations which combine a "free oscillation"

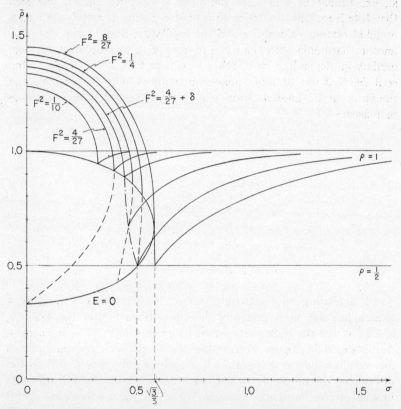

Fɪɢ. 17.1. Response curves for harmonic and combination oscillations for intermediate values of the detuning.

with a harmonic component to produce what is in general an almost-periodic response. However, for smaller values of σ and for values of F^2 near to but less than 8/27 the situation becomes more complicated:

a) For $1/4 < F^2 < 8/27$ the response curves for almost periodic oscillations continue to emerge from the line $\rho = 1/2$ and may for some values of F^2 cross the interior of the ellipse $E = 0$, as indicated.

For certain values of F^2 near to 8/27 one sees that two stable harmonic oscillations may exist (for the same value of F^2) for certain values of σ in the range $1/2 < \sigma < \sqrt{3}/3$, while for values of F^2 near to 1/4 there exists (again for the same F^2) a stable harmonic oscillation as well as a stable almost periodic oscillation. If one were to perform an experiment in which the detuning σ is slowly varied, one sees therefore from this discussion that sudden jumps in amplitude would occur at certain values of σ and the oscillation would jump either to another harmonic with the same frequency or to an almost periodic oscillation, depending on whether F^2 is near to 8/27 or to 1/4. One sees also that the jump phenomena are subject to hysteresis, since the jumps occur in different ways when σ is decreased than when it is increased.

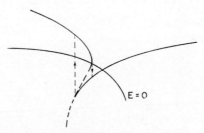

FIG. 17.2. Hysteresis phenomena.

b) According to Cartwright there exists a positive number δ (which has not been precisely determined) such that for $4/27 + \delta < F^2 < 1/4$ the response phenomena again exhibit curious features. As indicated in Figure 17.1 stable almost periodic response occurs for certain values of σ less than $1/2$ and these branches start in the interior of the ellipse $E = 0$ from a certain point on the unstable branch of the harmonic response curve for the given value of F^2. It follows that jump phenomena occur for this range of values of F^2 also, but now the transitions are always from a stable harmonic oscillation to a stable almost periodic oscillation or vice versa. One sees, too, that hysteresis occurs here also, with transitions occurring in the manner indicated in Figure 17.2.

c) If $F^2 < 4/27 + \delta$, the branch of the response curves referring to stable almost periodic oscillations has its origin always on the upper side of the ellipse $E = 0$ where the transition from stable to unstable harmonic oscillations occurs. To each value of σ there is only one point on the stable part of the response curve, and hence no jump phenomena occur in this range of values of F^2. For σ sufficiently small compared with F^2 we observe that the harmonic oscillation is the only stable oscillation, in accord with the result of Section 15.

Experimental confirmation of some of the results described here has been given. In particular, the transitions described in c) on the boundary of the ellipse have been observed. However, experimental confirmation of the "fine structure" of the response curves described in a) and b) above seems to be lacking, perhaps because the transitions from one type of phenomena to another occur for parameter values which are very close together; it may well be that the phenomena were just not noticed in making the experiments.

18. Subharmonic response

This chapter should not be closed without a reference to the important phenomena of subharmonic response in self-excited systems, which is important for the applications. That such subharmonics, i.e. oscillations whose least period is a fraction of the period of the excitation, occur is well known both experimentally and theoretically. In fact, subharmonics of order as high as the 200th are said to occur. It would be possible to treat these oscillations by the same kind of methods as we have used in Chapter IV to discuss the similar question for systems with nonlinear restoring forces, and also by the method of van der Pol used in the present chapter to treat the harmonic oscillations. We refrain from doing so, however, but note that this subject is treated at some length in the book of Minorsky [31].

CHAPTER VI

Hill's Equation and Its Application to the Study of the Stability of Nonlinear Oscillations

1. Mechanical and electrical problems leading to Hill's equation

A particle attached to one end of a light rigid rod which is pivoted at the other end is in equilibrium when the particle is either vertically above or below the pivot, but the former position—referred to as the inverted position of the pendulum—is obviously not a stable equilibrium position. If, however, the rod instead of being constrained to rotate about the fixed lower end is permitted to move so that the lower end slides freely on a vertical line, it is possible to convert the inverted unstable equilibrium position into a stable one by applying a vertical periodic force of proper amplitude and frequency to the lower end of the rod. We open the present chapter with a discussion of this problem (which is clearly of interest for its own sake) because it leads at once to the central question to be studied here in detail.

Fig. 1.1 indicates the rod and attached mass in a position displaced from the vertical. The motion is assumed to take place in the x, y-plane under the action of the weight mg, the external applied force $Y(t)$, and the force $X(t)$ provided by the constraint at the end A of the rod. The x-coordinate of point B (where the mass m is located) is given by

$$(1.1) \qquad x = l \sin \vartheta.$$

Since the mass of the rod is neglected the center of gravity of the system is at B and in addition the moment of inertia of the system with respect to this point is zero. The system has a plane motion which is determined by the equations of motion:

$$(1.2) \qquad m\ddot{x} = X, \qquad \text{and}$$

$$(1.3) \qquad Yl \sin \vartheta - Xl \cos \vartheta = 0.$$

189

Equation (1.2) results from the principle of the motion of the center of gravity, and (1.3) results from the angular momentum law, in view of the fact that the moment of inertia about the center of gravity B is zero. Equations (1.1), (1.2), and (1.3) would serve to determine the motion.

We assume now that the angle ϑ is so small that we may replace $\sin \vartheta$ by ϑ and $\cos \vartheta$ by unity, and eliminate x and X from (1.2) by using (1.1) and (1.3); the result is the following differential equation for the angle ϑ:

$$(1.4) \qquad ml\ddot{\vartheta} - Y\vartheta = 0.$$

The applied vertical force $Y(t)$ is assumed to be given as follows:

$$(1.5) \qquad Y(t) = mg - mp(t);$$

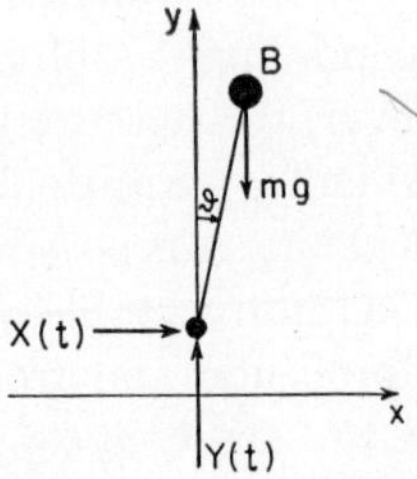

FIG. 1.1. Pendulum with a prescribed vertical force at one end.

that is, it consists of a constant part mg equal to the static weight to be supported and in addition a variable part depending on the time. In this case (1.4) becomes

$$(1.6) \qquad \ddot{\theta} + \left(-\frac{g}{l} + \frac{1}{l}\,p(t)\right)\theta = 0.$$

If the function $p(t)$ is periodic in t—which we assume from now on—*the linear equation (1.6) is called Hill's equation.*

In deriving (1.6) we assumed the angle θ to be small, that is, we tacitly assumed that the inverted pendulum could be made stable by a proper choice of the periodic part $-mp(t)$ of the vertical force $Y(t)$. It is clear that θ could not be expected to remain small for small but otherwise arbitrary initial values for θ and $\dot{\theta}$ unless *all solutions of (1.6) are bounded for all positive values of t.* We note that if $p(t) \equiv 0$ the solutions of (1.6) are linear combinations of $e^{\sqrt{g/l}\,t}$ and $e^{-\sqrt{g/l}\,t}$ and hence have unbounded solutions among them, and this corresponds to the fact that the inverted pendulum is unstable unless some force is supplied in addition to what is needed to support the static weight. It is far from obvious that $p(t)$ can be chosen in such a way that all solutions of (1.6) are bounded. However, as we shall see later, *one may take $p(t)$ in the form $p(t) = A \cos \omega t$ and dispose of the constants A and ω* (even for arbitrarily small values of A) *in such a way that (1.6) has only bounded solutions for all t.*

In other words, the unstable inverted position of equilibrium of the pendulum can be converted into a stable one if a vertical pulsating force of proper amplitude and frequency is applied at the support of the pendulum. It is also of interest to add that it is possible to convert the stable equilibrium position of the pendulum (that is, the normal position in which the mass lies below the support) into an unstable one by applying a properly chosen periodic vertical force at the support. The differential equation for this latter case is the same as (1.6) but with the sign of g reversed, corresponding to a reversal in the direction of the force of gravity. If we take $p(t) \equiv 0$ we see that all solutions of the resulting differential equation are bounded—they yield in fact the simple harmonic motions given by linear combinations of $\cos \sqrt{\dfrac{g}{l}}\,t$ and $\sin \sqrt{\dfrac{g}{l}}\,t$. We shall see later that periodic forces $mp(t)$ of arbitrarily small amplitude can be chosen in such a way that the stable "normal" position of the pendulum is made unstable.

A simple mechanical-electrical system leading to a Hill's equation is also readily devised. It consists of a circuit containing a constant inductance and a condenser in series. The plates of the condenser are assumed to be movable and to be actuated mechanically in such a way as to change the capacity of the condenser periodically in the time. If q is the charge on the condenser, L the inductance, and $C(t)$ the capacitance of the condenser, we have for q the differential equation

$$(1.7) \qquad L \frac{d^2q}{dt^2} + \frac{1}{C(t)}\, q = 0,$$

and since $C(t)$ is assumed to be periodic in t (1.7) is a Hill's equation. Such a mechanical-electrical system is a convenient one for experimental investigations of Hill's equation (cf., for example, Barrow [4]).

One could easily give many other examples of physical problems which lead to a Hill's equation (cf., for example, the first chapter of the book of Strutt [37]).

The main reason for our interest in this particular *linear* equation in a book devoted to nonlinear vibrating systems is that *the problem of the infinitesimal stability of the periodic solutions of our nonlinear systems always leads to a Hill's equation*, as we have seen (cf., e.g., Chapter IV, Section 11 for the case of the Duffing equation). In fact, the Hill's equation in these cases is a variational equation characterizing small variations from the given periodic motion whose stability is to be investigated. *We say that a given periodic motion is stable if all solutions of the variational equation* (which is always a Hill's equation in our cases) *associated with it are bounded for all positive values of t, and unstable if the variational equation has an unbounded solution.* In other words, the question of stability of a given motion depends upon the character of the totality of solutions of the Hill's equation associated with the motion, just as in the problem of the inverted pendulum discussed above.* Thus it makes no sense to speak of a stable solution of the Hill's equation; but it is nevertheless very convenient to use such a phrase as "the solutions are stable" as a concise way of saying that all solutions of the given Hill's equation are bounded. We shall frequently make use of such a terminology when no confusion is likely to arise because of it.

The remainder of this chapter is devoted to a discussion of the general theory of Hill's equation with particular reference to the question of boundedness of the solutions. The important special case of the Mathieu equation—a Hill's equation in which the periodic coefficient is a simple harmonic function of the independent variable— is taken up in some detail in order to open the way to a discussion of the stability of the periodic solutions of Duffing's equation obtained in Chapter IV.

The author cannot forbear to add that the theory of Hill's equation is a theory of extraordinary elegance and one that is well worth studying on its own merits, quite aside from its utility in discussing stability questions.

---

* The Hill's equation (1.6) is also a variational equation for the inverted pendulum problem: the quantity θ could really have been written $\delta\theta$ and interpreted as a small variation of the basic motion given by $\theta \equiv 0$.

2. Floquet theory for linear differential equations with periodic coefficients

We consider differential equations of the form

$$(2.1) \qquad \frac{d^2w}{dz^2} + p(z)\,\frac{dw}{dz} + q(z)w \equiv 0$$

in which the coefficients $p(z)$ and $q(z)$ are regular periodic functions of z of real period Ω, i.e. $p(z + \Omega) = p(z)$, $q(z + \Omega) = q(z)$. In this section it is convenient (though not necessary) to consider p and q as analytic functions of the complex variable z defined in a strip containing the entire real axis.

Before drawing any conclusions resulting from the periodicity of p and q it is useful to recall a few well known results about the solutions $w(z)$ of (2.1) which follow solely from the fact that (2.1) is linear and homogeneous with coefficients which are regular functions of z. First and foremost, there exists a pair of linearly independent and regular solutions $w_1(z)$ and $w_2(z)$ neither of which vanishes identically, and every other solution w is a linear combination of these two:

$$(2.2) \qquad w = c_1w_1 + c_2w_2 .$$

The pair of solutions w_1, w_2 is called a *fundamental set* of solutions.

The necessary and sufficient condition that w_1 and w_2 form a fundamental set is that the Wronskian determinant

$$(2.3) \qquad \Delta(z) = \begin{vmatrix} w_1 & w_2 \\ \dfrac{dw_1}{dz} & \dfrac{dw_2}{dz} \end{vmatrix} \equiv \begin{vmatrix} w_1 & w_2 \\ w_1' & w_2' \end{vmatrix}$$

should not vanish identically in z. (In this chapter the primed symbols refer to differentiations with respect to z.) This is readily proved as follows: If non-vanishing constants c_1, c_2 exist such that $c_1w_1 + c_2w_2 \equiv 0$, it follows that $c_1w_1' + c_2w_2'$ also vanishes identically so that Δ vanishes identically in this case. Hence if Δ does not vanish it follows that w_1 and w_2 are not linearly dependent. The non-vanishing of Δ is thus a sufficient condition for the linear independence of w_1 and w_2. On the other hand, if $\Delta = w_1w_2' - w_2w_1' \equiv 0$, we can integrate to find $\log \dfrac{w_1}{w_2} = $ constant so that w_1 and w_2 are necessarily linearly dependent. Thus $\Delta \neq 0$ is both a necessary and a sufficient condition for the linear independence of w_1 and w_2.

It is of interest also to note that if $\Delta(z)$ vanishes for a single value of z then it vanishes for all z. This follows from a formula for $\Delta(z)$ which will be useful to us later on. The formula is obtained in the following way: Differentiation of both sides of (2.3) with respect to z yields

$$\frac{d\Delta}{dz} = \begin{vmatrix} w_1 & w_2 \\ w_1'' & w_2'' \end{vmatrix} + \begin{vmatrix} w_1' & w_2' \\ w_1' & w_2' \end{vmatrix},$$

from which we obtain

$$(2.4) \qquad \frac{d\Delta}{dz} = \begin{vmatrix} w_1 & w_2 \\ w_1'' & w_2'' \end{vmatrix}.$$

We now make use of the fact that w_1 and w_2 are both solutions of (2.1) and accordingly replace w_1'' and w_2'' in (2.4) by their values as given by (2.1). The result, after expanding the determinant in (2.4) as a sum of two determinants and noting that one of the latter vanishes, is as follows:

$$(2.5) \qquad \frac{d\Delta}{dz} = -p\Delta.$$

This linear differential equation for Δ has the general solution

$$(2.6) \qquad \Delta(z) = \Delta_0 \exp\left[-\int_{z_0}^{z} p(\xi)\, d\xi \right],$$

in which Δ_0 is the value of Δ for $z = z_0$. Since $p(z)$ is regular for all z it follows that the exponential function in (2.6) never vanishes and hence Δ cannot vanish at any point without vanishing identically.

We turn now to a discussion of some of the consequences which follow from the assumed periodicity of the coefficients p and q.* To begin with, if $w_1(z)$ and $w_2(z)$ form a fundamental set of solutions it follows that $w_1(z + \Omega)$ and $w_2(z + \Omega)$ also form a fundamental set since they clearly satisfy (2.1) because of the fact that $p(z + \Omega) = p(z)$ and $q(z + \Omega) = q(z)$ and the fact that $\Delta(z + \Omega)$ does not vanish, as we have just seen. Since every solution can be obtained as a

---

* This theory is usually called the Floquet theory.

linear combination of w_1 and w_2 it follows that the solutions $W_1(z) = w_1(z + \Omega)$ and $W_2(z) = w_2(z + \Omega)$ satisfy linear relations of the form

$$(2.7) \qquad \begin{cases} W_1(z) = w_1(z + \Omega) = a_{11}w_1(z) + a_{12}w_2(z), \\ W_2(z) = w_2(z + \Omega) = a_{21}w_1(z) + a_{22}w_2(z). \end{cases}$$

The Wronskian of W_1, W_2 is readily seen to result from the Wronskian of w_1, w_2 through multiplication by the determinant of the coefficients in (2.7); that is, we have

$$(2.8) \qquad \Delta(z + \Omega) = \begin{vmatrix} a_{11} & a_{12} \\ a_{21} & a_{22} \end{vmatrix} \cdot \Delta(z).$$

Since W_1 and W_2 form a fundamental set it follows therefore that

$$\begin{vmatrix} a_{11} & a_{12} \\ a_{21} & a_{22} \end{vmatrix} \neq 0.$$

It is not true in general that the solutions $w(z)$ of (2.1) are periodic with real periods in z though the coefficients have a real period, as one sees in the special case $\dfrac{d^2w}{dz^2} - w = 0$ in which all solutions are linear combinations of exponentials with pure imaginary periods. There are, however, solutions which have the property that they are multiplied by a constant factor when z is shifted by the amount of the period Ω (assumed always to be real), that is, there exist solutions $W(z)$ for which

$$(2.9) \qquad W(z) = w(z + \Omega) = \sigma w(z), \qquad \sigma = \text{constant},$$

for all z. Such solutions are called *normal solutions*; they play a central role in the discussion to follow for reasons which will soon be apparent. Any normal solution w (assuming that it exists) can be expressed as an appropriate linear combination of w_1 and w_2 :

$$(2.10) \qquad w = \lambda_1 w_1 + \lambda_2 w_2 .$$

A normal solution w satisfies (2.9) while w_1 and w_2 satisfy (2.7); it follows that the relation

$$[\lambda_1(a_{11} - \sigma) + \lambda_2 a_{21}]w_1 + [\lambda_1 a_{12} + \lambda_2(a_{22} - \sigma)]w_2 = 0$$

holds identically in z, and since w_1 and w_2 are linearly independent their coefficients must vanish. Furthermore, since the constants λ_1 and

λ_2 do not both vanish, it follows that the following determinantal equation, called the *characteristic equation*:

$$(2.11) \qquad \begin{vmatrix} a_{11} - \sigma & a_{21} \\ a_{12} & a_{22} - \sigma \end{vmatrix} = 0,$$

holds.* The equation (2.11) is a quadratic equation for σ neither root of which vanishes since the constant term in the equation is $\begin{vmatrix} a_{11} & a_{21} \\ a_{12} & a_{22} \end{vmatrix}$ and we have already seen (cf. (2.8) and the remarks following it) that this determinant does not vanish. We must consider two cases according to whether the roots of (2.11) are unequal or not.

If the roots σ_1, σ_2 of the characteristic equation are unequal there exists a pair of linearly independent normal solutions. This follows from the fact that each of the roots σ_1 and σ_2 yields a pair of values of λ_1 and λ_2 which in turn yield a normal solution, and since these two pairs of values of λ_1 and λ_2 are themselves linearly independent since σ_1 and σ_2 are different it follows that the two normal solutions thus obtained are linearly independent. *The two normal solutions in this case can be shown to consist of the product of an exponential function and a periodic function of period Ω.* For this purpose consider the normal solution w_i corresponding to σ_i. We may write

$$e^{-\alpha_i(z+\Omega)} w_i(z + \Omega) = \sigma_i e^{-\alpha_i \Omega} e^{-\alpha_i z} w_i(z),$$

and hence the function $\varphi_i(z)$ defined by $\varphi_i(z) = e^{-\alpha_i z} w_i(z)$ is seen to be a periodic function with period Ω if α_i is chosen so that

$$(2.12) \qquad e^{\alpha_i \Omega} = \sigma_i.$$

It follows that the linearly independent normal solutions w_1 and w_2 can be expressed in the form

$$(2.13) \qquad \begin{cases} w_1(z) = e^{\alpha_1 z} \varphi_1(z), \\ w_2(z) = e^{\alpha_2 z} \varphi_2(z), \end{cases}$$

in which $\alpha_1 \neq \alpha_2$ and φ_1 and φ_2 both have the period Ω. It should be noted that σ_1 and σ_2, and also α_1 and α_2, need not be real numbers.

---

* It is not difficult to show that the equation for σ is the same no matter what pair of fundamental solutions one takes as a basis for the above developments.

Later on we shall be particularly interested in the cases in which the quantities σ_i have the values ± 1. If $\sigma_i = +1$, for example, we note from the definition of the normal solutions that the *corresponding normal solution w_i is periodic of period Ω, while if $\sigma_i = -1$ the normal solution w_i has the period 2Ω*; at the same time the quantity $\alpha_i \Omega$ in (2.12) must have its real part zero in both cases and its imaginary part equal to zero for $\sigma_i = +1$ and equal to $i\pi$ for $\sigma_i = -1$. It is also of interest to observe that normal solutions with the smallest period $k\Omega$, with k any integer, will occur if $\sigma_i = e^{2\pi i/k}$.

For later purposes it is of importance to consider the exceptional case in which (2.11) has repeated roots, i.e., $\sigma_1 = \sigma_2 = \sigma$. There is then at least one normal solution w_1 which behaves like those in the preceding case. As we shall see, any other linearly independent solution behaves in a different fashion in general. To study this question, we introduce the normal solution w_1 itself as one of a pair of fundamental solutions and choose the other fundamental solution w_2 arbitrarily. In this case the linear substitution (2.7) has the special form

$$\begin{cases} W_1(z) = \sigma w_1(z), \\ W_2(z) = a w_1(z) + b w_2(z). \end{cases}$$

Since, however, (2.11) is assumed to have a double root it is clear that $b = \sigma$. It follows that the quotient $\dfrac{w_2}{w_1}$ undergoes the substitution

$$\frac{W_2}{W_1} = \frac{w_2}{w_1} + \frac{a}{\sigma}$$

when z is shifted by the amount Ω, and hence that the difference

$$\frac{w_2}{w_1} - \frac{a}{\sigma}\frac{z}{\Omega}$$

is a periodic function of period Ω, so that w_2 is of the form

$$w_1\left[\frac{a}{\sigma}\frac{z}{\Omega} + \psi(z)\right]$$

with $\psi(z)$ a function with period Ω. Thus when the characteristic equation has repeated roots there exist two fundamental solutions w_1, w_2 of the form

$$(2.14) \qquad \begin{cases} w_1(z) = e^{\alpha z}\, \varphi_1(z), \\ w_2(z) = e^{\alpha z} \left[\dfrac{az}{o\Omega} \cdot \varphi_1(z) + \psi_1(z) \right], \end{cases}$$

in which $\varphi_1(z)$ and $\psi_1(z)$ are periodic functions of period Ω.

As we have remarked earlier, our principal interest is in the question whether all solutions of (2.1) are bounded for all real positive z or not. If all solutions of (2.1) are bounded, we should say, in view of the discussion in the preceding section, that the *differential equation* (2.1) characterizes a stable situation or motion, and otherwise an unstable situation; but to avoid such a lengthy terminology we shall in general say that *the solutions of* (2.1) *are stable if all are bounded and unstable if an unbounded solution exists.*

On the basis of (2.13) we see that this question is settled by the values of the constants α_1 and α_2 alone in case the roots σ_1 and σ_2 are unequal: all solutions are bounded if, and only if, $|e^{\alpha_i z}|$ is bounded for all real positive z and this in turn requires that α_1 and α_2 should have real parts which are not positive. In the case of repeated roots of the characteristic equation we see from (2.14) that if α has a real part which is negative all solutions of (2.1) are bounded. If, however, the real part of α is zero the solutions are stable in the case of repeated roots of the characteristic equation only if in addition the constant a in (2.14) is zero, which means that two linearly independent normal solutions would exist. This special situation, in which the real part of α is zero and the roots σ_1 and σ_2 are equal, is of considerable importance in the applications to be treated later.

3. The stability problem for Hill's equation and the Mathieu equation.*

The Floquet theory, as outlined in the preceding section, reveals the functional character of the solutions of Hill's equation, but does not provide a means of deciding the stability question. This problem is, in fact, a difficult one which can be solved only by studying the

---

* In writing this section we have been greatly aided by the book of Strutt [37], which should be consulted for further details and for extensive references to the literature.

solutions of the given differential equation in considerable detail. For the special case of Hill's equation:

$$(3.1) \qquad \frac{d^2 w}{dz^2} + q(z) \cdot w = 0,$$

i.e. the special case of (2.1) in which $p(z) = 0$, *and in which also we restrict ourselves to functions of the real variable z*, it is possible to carry the analysis of the preceding section somewhat farther. Such an extension of the Floquet theory will be carried out in the present section.

We begin by choosing as fundamental solutions w_1, w_2 of (3.1) the pair of solutions which satisfy the following initial conditions* at $z = 0$:

$$(3.2) \qquad \begin{cases} w_1(0) = 1, & w_1'(0) = 0 \\ w_2(0) = 0, & w_2'(0) = 1. \end{cases}$$

That such a pair of solutions is linearly independent is clear, since the Wronskian is different from zero for $z = 0$—it has, in fact, the value one. With this pair of functions as a basis we follow the method of the preceding section to determine a normal solution. Upon shifting by the amount of the real period Ω the functions w_1 and w_2 become

$$(3.3) \qquad \begin{cases} w_1(z + \Omega) = a_{11}w_1 + a_{12}w_2, \\ w_2(z + \Omega) = a_{21}w_1 + a_{22}w_2. \end{cases}$$

The coefficients a_{ik} are real in the present case since w_1 and w_2 are real. For a normal solution $w(z)$ we have $w(z + \Omega) = \sigma w(z)$ with σ a root of

$$\begin{vmatrix} a_{11} - \sigma & a_{21} \\ a_{12} & a_{22} - \sigma \end{vmatrix} = 0,$$

which, as we shall prove presently, can be written in the form

$$(3.4) \qquad \sigma^2 - A\sigma + 1 = 0,$$

with A obviously a real number.**

---

* We assume as known the fact that solutions of (3.1) exist for any prescribed initial conditions and for all values of z.

** The roots of (3.4) need not and in general will not be real. Consequently the normal solutions will in general be complex-valued functions of the real variable z.

We observe that the constant term in (3.4) is in any case the determinant $\begin{vmatrix} a_{11} & a_{21} \\ a_{12} & a_{22} \end{vmatrix}$ and we are claiming that it has the value one. This follows from the fact that the Wronskian for any pair of fundamental solutions of (3.1) has a constant value for all values of z, as we see from (2.6) since $p(z) \equiv 0$ in the present case. The Wronskian $\Delta(z + \Omega)$ of $w_1(z + \Omega)$ and $w_2(z + \Omega)$ therefore has the same value as the Wronskian $\Delta(z)$ for $w_1(z)$ and $w_2(z)$ since the former pair of functions continues to be a fundamental set of solutions, and hence $\Delta(z + \Omega)$ has the value one because of (3.2). But the Wronskian of $w_1(z + \Omega)$ and $w_2(z + \Omega)$ is the product of $\begin{vmatrix} a_{11} & a_{12} \\ a_{21} & a_{22} \end{vmatrix}$ and the Wronskian of $w_1(z)$ and $w_2(z)$ (cf. (2.8)), and our result follows at once.

In the present case, in which the differential equation does not contain a term in the first derivative $\dfrac{dw}{dz}$, the important relation

$$(3.5) \qquad\qquad \sigma_1 \sigma_2 = 1$$

therefore holds between the two roots σ_1, σ_2 of the characteristic equation belonging to the normal solutions. It is useful to add the remark that in the case of repeated roots of the characteristic equation $\sigma_1 = \sigma_2 = \sigma$ and σ must have one or the other of the values ± 1; thus the equation (3.1) necessarily has a periodic solution in this case.

For stability it is obviously necessary, because of the definition of the normal solutions, to require that $|\sigma_1|$ and $|\sigma_2|$ should both satisfy the inequality $|\sigma_1| \leq 1$, and this in view of (3.5) means that σ_1 and σ_2 must satisfy the relations

$$(3.6) \qquad\qquad |\sigma_1| = |\sigma_2| = 1.$$

The relations (3.6) are thus necessary conditions for stability in the present cases; they are also sufficient conditions for stability in case $\sigma_1 \neq \sigma_2$, as we know from the form of the normal solutions given by (2.13), which is valid in this case, and the fact that these two solutions form a fundamental set. In case $\sigma_1 = \sigma_2$, the conditions (3.6) are not sufficient for stability; we must in fact require in addition that the constant a in (2.14) should vanish.

In the problems leading to Hill's equation discussed at the beginning of this chapter the periodic coefficient corresponding to $q(z)$ in (3.1) contained a number of parameters, and the stability question referred to the proper choice of the parameters in order to insure that all solutions be stable. The parameter values at which a transition from stability to instability occurs are then of particular interest; we propose to discuss such transitions in the important special case for which (3.1) takes the form

$$(3.7) \qquad \frac{d^2w}{dz^2} + (\delta + \epsilon r(z))w = 0.$$

The parameters are δ and ϵ, and the function $r(z)$ is assumed to have the average value zero over its period Ω. We turn now to the characteristic equation (3.4) and observe that once $r(z)$ has been fixed the real number A is a function of δ and ϵ only. The roots of this equation are given by

$$(3.8) \qquad \sigma_{1,2} = \frac{A}{2} \pm \sqrt{\left(\frac{A}{2}\right)^2 - 1}.$$

In view of the stability condition (3.6) and the fact that A is real we see that

a) If $|A| > 2$, the solutions are clearly unstable, and

b) If $|A| < 2$, the solutions are stable. This follows because $\sigma_1 \neq \sigma_2$, $|\sigma_1| = |\sigma_2| = 1$ and the α_i in (2.13) are therefore pure imaginary (cf. 2.12). Since A is real it follows that the transition from stability to instability occurs for $A = +2$, or $A = -2$, which corresponds to the repeated roots $\sigma = +1$, or $\sigma = -1$, and hence also to the existence of a periodic normal solution of period Ω or 2Ω.

If the quantities δ, ϵ are such that the solutions are stable we call such values of δ and ϵ *stable values*, and otherwise *unstable values*. Values of δ and ϵ for which $\sigma = \pm 1$ are called *transition values*. A normal solution $w(z)$ is characterized by the property $w(z + \Omega) = \sigma w(z)$ and hence we can say that *corresponding to transition values of δ and ϵ from stability to instability there must exist at least one periodic solution of (3.7) with the period Ω or the period 2Ω*. The transition values themselves usually lead to unstable solutions and hence belong themselves to the unstable values, in view of the remarks at the end of the preceding section, since the characteristic equation has repeated roots of absolute value one in this case.

The transition values of δ and ϵ satisfy the equations

$$(3.9) \qquad \begin{cases} A(\delta, \epsilon) = 2 \\ A(\delta, \epsilon) = -2. \end{cases}$$

It has been shown by O. Haupt [41]* that the points satisfying (3.9)
fill out curves in a δ, ϵ-plane separating it into regions in which
(δ, ϵ) have stable or unstable values. In fact, these regions can be
described in more detail, as follows: For each fixed ϵ there exists an
infinite set δ_i of isolated values of δ bounded on the negative (left)
side of the δ-axis but unbounded on the positive side that satisfy
(3.9). Upon moving from left to right along the δ-axis the points δ_i
fall into pairs of adjacent points (with the exception of the first
point at the extreme left) in such a way that one pair satisfies the
first of (3.9) while the next following pair satisfies the second of (3.9).
This means also that the periodic solutions of (3.7) correlated with
these transition values of δ (for fixed ϵ, we recall) are also arranged in
successive pairs which have alternately the periods Ω and 2Ω, except
for the first point on the left to which a solution of period Ω is correlated followed by two of period 2Ω at the next two points. Furthermore, O. Haupt shows that all values of δ between pairs of points
of the same type yield unstable solutions while the other values of
δ yield stable solutions. The points from $-\infty$ up to the first point
δ_0 on the δ-axis yield unstable solutions. The transition points δ_i
themselves belong to the unstable regions in general, as we know
from the discussion following (3.8), since the characteristic equation
has a double root in such cases. The general situation is indicated
schematically in the accompanying figure.

4. The Mathieu equation

If we take $r(z) \equiv \cos z$ in (3.7) the result is a special case of Hill's
equation called the Mathieu equation:

$$(4.1) \qquad \frac{d^2w}{dz^2} + (\delta + \epsilon \cos z)w = 0.$$

* The theorems of Haupt make extensive use of the theory of linear eigen-
value problems. We do not reproduce these proofs here. However, we are
interested only in applying such a theory for the case of small values of ϵ,
and this can be done with sufficient accuracy in a rather simple way, as we
shall see in Section 5.

This equation has been studied very extensively, and, in particular, the stable and unstable regions in the δ, ϵ-plane have been determined completely for all values of δ and ϵ. In the following section we shall determine these regions approximately for small values of ϵ as a basis for a study of the stability of the harmonic solutions of the Duffing equation, which is to be carried out in sec. 6). Such a determination of the stability regions even for small values of ϵ requires, however, a knowledge of certain facts about the solutions of (4.1), and, since these facts are of general interest in any case—for one thing because the Mathieu equation governs a wide variety of problems in mathematical physics—we propose to investigate them in the present section.

We begin our study of the properties of the solutions of (4.1) by remarking that if $w(z)$ is a solution that is neither an even nor an odd function of z, then $w(-z)$ is obviously also a solution and $w(z)$ and $w(-z)$ form a fundamental set. Thus $[w(z) + w(-z)]$ would be an even solution and $[w(z) - w(-z)]$ an odd solution and neither vanishes identically. On the other hand (4.1) cannot possess two linearly independent even solutions since otherwise no solution satisfying the initial conditions $w(0) = 0$, $w_0'(0) = 1$ would exist; in the same way one sees that no two distinct odd solutions exist. We conclude that *the Mathieu equation possesses a fundamental set of solutions w_1, w_2 in which one solution is an even and the other an odd function of z.*

We are particularly interested in the transition values of δ and ϵ from stable to unstable values. From the discussion of the preceding section we know that (4.1) possesses for such values of δ and ϵ a periodic solution having either the period 2π or the period 4π, since the periodic coefficient has the period 2π in this case. Since these periodic solutions are regular for all values of z it follows that they possess Fourier series developments. Thus in case the solution $w(z)$ has the period 2π it would have a development of the form

$$(4.2) \qquad w(z) = a_0 + \sum_{n=1}^{\infty} (a_n \cos nz + b_n \sin nz).$$

If one inserts this series in (4.1) it turns out, however, that if the first non-vanishing term is for example a cosine term then all subsequent terms are also cosines, and likewise if the first non-vanishing term is a sine term then all subsequent terms are sines. It follows therefore that the Fourier series for a solution of period 2π is either a cosine series or a sine series:

(4.3)
$$
\begin{cases}
w(z) = a_0 + \sum_{n=1}^{\infty} a_n \cos nz, & \text{or} \\
w(z) = \sum_{n=1}^{\infty} b_n \sin nz.
\end{cases}
$$

A corresponding statement holds for the periodic solutions of period 4π. A little later we shall show that in general only one of the two possibilities can actually occur for a given pair of transition values of ϵ and δ—in other words there can be only one periodic solution in such a case (at least within a constant multiplying factor). This fact makes it possible to calculate the transition values in the following manner. Insertion of the series (4.3) in (4.1) leads to one or the other of the following sets of recurrence relations for the a_n and b_n, as one can readily verify:

(4.4)
$$
\begin{cases}
\delta a_0 + \dfrac{\epsilon}{2} a_1 = 0, \\
(\delta - n^2)a_n + \dfrac{\epsilon}{2}(a_{n-1} + a_{n+1}) = 0, & n = 1, 2, \cdots
\end{cases}
$$

(4.5)
$$
\begin{cases}
(\delta - 1)b_1 + \dfrac{\epsilon}{2} b_2 = 0, \\
(\delta - n^2)b_n + \dfrac{\epsilon}{2}(b_{n-1} + b_{n+1}) = 0, & n = 2, 3, \cdots.
\end{cases}
$$

Suppose we content ourselves with the accuracy obtained by taking a certain finite number of terms of (4.3). The relations (4.4) and (4.5) then furnish two distinct sets of linear homogeneous equations which must be satisfied for values of a_n and for values of b_n which do not all vanish. The determinants of the coefficients—often referred to as Hill's determinants—must therefore vanish, and these equations between ϵ and δ furnish approximations to the transition values of ϵ and δ. In a similar fashion the transition values corresponding to the periodic solutions of period 4π can be determined.

In this way the stability regions of the Mathieu equation have been completely determined; we reproduce the results in Fig. 4.1 in which the shaded regions are the stable regions. These regions are shown only for $\epsilon \geq 0$; for $\epsilon < 0$ the stable regions are obtained by reflection in the δ-axis.* The stable regions are connected together at the points $\delta = \dfrac{n^2}{4}$, $\epsilon = 0$, n an integer; for these values of δ and ϵ the equation (4.1) obviously possesses the bounded solutions $\cos \dfrac{n}{2} z$ and $\sin \dfrac{n}{2} z$ which form a fundamental set of period 2π if n is even and

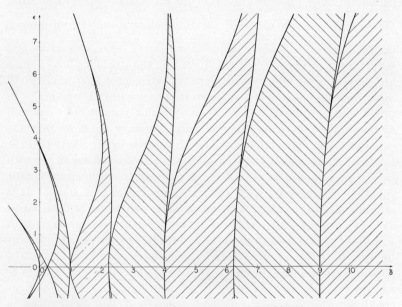

FIG. 4.1. Stable and unstable regions for the Mathieu equation.

4π if n is odd. It has been shown that for large values of ϵ the stable regions become very narrow and tend to curves having the slope -1, as the figure indicates. For negative values of δ the stable regions are quite narrow.

We are now in a position to settle the question raised in section 1) about the stability of the inverted pendulum. If we set $\dfrac{1}{l} \, p(t) = \epsilon \cos t$ and $-\dfrac{g}{l} = \delta$ in equation (1.6) we have a Mathieu equation

* In the next section we shall determine these regions approximately for small values of ϵ and for $0 \leq \delta \leq 1$.

with a negative value for δ. It follows from Fig. 4.1 that values for ϵ (that is, values for the amplitude of the vibratory force at the support) can really be found for which (1.6) has only stable solutions, since stable regions of the $\delta\epsilon$-plane exist for negative values of δ. We observe also that the stable equilibrium position of the pendulum, for which $\dfrac{g}{l} = \delta$ is positive, can be made unstable by a suitable choice of the amplitude of the pulsating force at the support of the pendulum, since unstable regions of the δ, ϵ-plane occur for δ positive.

We have already observed that the points $\delta = \dfrac{n^2}{4}$, n an integer, on the δ-axis are boundary points of the stable regions (or as we also say, transition points) which also belong to the stable regions. These are, however, *the only transition points which have this property:* all other boundary points of the stable regions are unstable points. To prove this fact, which will be of considerable importance in certain of our later investigations, it is only necessary to show that the equation (4.1) cannot have a pair of linearly independent periodic solutions since in that case the solutions are of the form (2.14) with $\alpha = 0$ or πi and $a \neq 0$ because of the fact that the characteristic equation has repeated roots for transition values of δ and ϵ. It is not difficult to prove that only one periodic solution of (4.1) of period 2π or 4π exists for all transition values of δ and ϵ except those with $\epsilon = 0$. This has been done by E. L. Ince [17] by an indirect proof as follows: If there were two linearly independent periodic solutions of period 2π, for example, we know that we could choose one of them as an odd function and the other as an even function; these solutions could be expressed as the Fourier cosine and sine series given in (4.3). The coefficients of these series would then satisfy the recurrence relations (4.4) and (4.5). By eliminating $(\delta - 1)$ from the equation for $n = 1$ in (4.4) and the first equation in (4.5) we obtain the relation

$$a_0 \, b_1 = \begin{vmatrix} a_1 & a_2 \\ b_1 & b_2 \end{vmatrix}$$

provided that $\epsilon \neq 0$. By using the equations for $n = r$ in both groups one obtains in similar fashion

$$\begin{vmatrix} a_r & a_{r+1} \\ b_r & b_{r+1} \end{vmatrix} = \begin{vmatrix} a_{r-1} & a_r \\ b_{r-1} & b_r \end{vmatrix}$$

so that we have the following relation, valid for all $n = 1, 2, \cdots$:

(4.6)
$$\begin{vmatrix} a_n & a_{n+1} \\ b_n & b_{n+1} \end{vmatrix} = a_0 b_1 .$$

We now observe that if $a_0 = 0$, with $\epsilon \neq 0$, it follows from (4.4) that all a_n are zero, and similarly that all b_n vanish if $b_1 = 0$, $\epsilon \neq 0$. Therefore $a_0 \neq 0$ and also $b_1 \neq 0$ since the two solutions are assumed to be a fundamental set, and it follows that $a_0 b_1$ is a non-zero constant if $\epsilon \neq 0$. On the other hand, the existence of the solutions (4.3) requires that the Fourier series converge and hence that $a_n \to 0$ and $b_n \to 0$ as $n \to \infty$, and this is obviously not compatible with (4.6). It follows that two periodic solutions of period 2π of the Mathieu equation cannot exist for transition values of ϵ and δ except when $\epsilon = 0$, in which case $\delta = n^2$. In the same way it can be shown that two periodic solutions of period 4π cannot exist unless $\epsilon = 0$, in which case $\delta = \dfrac{n^2}{4}$, n odd. At boundary points of the stability region not on the δ-axis it follows therefore that all solutions cannot be periodic and consequently that unbounded solutions exist which tend to infinity like the first power of z.

We close this section with a few remarks about periodic solutions of (4.1) in general. We have seen that for values of δ and ϵ corresponding to certain boundary points of the stable regions there exist solutions of period 2Ω and for other boundary points solutions of period Ω, in which Ω is the period of the periodic coefficient. Since, as we have seen in the first section of this chapter, the periodic coefficient corresponds in general to a periodic disturbance having its origin outside the system we see that periodic solutions of (4.1) may exist which have the same frequency or which have half the frequency of the disturbance. In other words, we might say that the subharmonic solution of order $\frac{1}{2}$ occurs for appropriate choices of ϵ and δ as well as the harmonic solution. In the engineering literature where this fact is of importance (in the theory of the whirling of shafts, for example) it is often overlooked that still other types of periodic solutions of Hill's equation and the Mathieu equation may occur. The fact is that all of the types of periodic solutions discussed in earlier chapters—the subharmonics of all orders, as well as the ultra-harmonics and ultra-subharmonics—occur in these cases also when ϵ and δ are properly chosen values *taken from the stable regions* of the δ, ϵ-*plane*.   It is not difficult to see in a general way how this comes about, as follows: For values of ϵ and δ in the stable regions as we have seen in section 2) (cf. equation (2.13)), the normal solutions have the form

$$(4.7) \qquad\qquad w(z) = e^{\alpha z}\varphi(z)$$

with α a pure imaginary constant and $\varphi(z)$ a function of period Ω. It can be shown that there exist normal solutions for which

$$i\alpha = \frac{2\pi}{\Omega} \cdot \frac{p}{q}$$

with p and q any integers prime to each other.   If $p/q = 1$ or $\frac{1}{2}$ we have the transition cases corresponding to boundary points of the stable regions and to periodic solutions having the periods Ω or 2Ω. For any other values of p/q *all* solutions $w(z)$ corresponding to stable values of δ and ϵ are periodic of period $q\Omega$ since two linearly independent normal solutions of the form (3.15) exist with conjugate complex values of α, as we know.   The cases $p = 1, q > 1$ correspond to the subharmonics, the cases $q = 1, p > 1$ to the ultra-harmonics, and the cases $q > 1, p > 1$ to the ultra-subharmonics.   The values of δ and ϵ corresponding to solutions of each of these types fill out curves in the δ, ϵ-plane which lie in the stable regions and which presumably have much the same appearance as the boundary curves of the stable regions which correspond to the occurrence of the harmonic solutions and the subharmonics of order $\frac{1}{2}$.

5. *Stability of the solutions of the Mathieu equation for small values of* ϵ

For the purpose of discussing the stability of the periodic solutions of a nonlinear system (with one degree of freedom) in the neighborhood of the linearized system it is usually sufficient to solve the stability problem for a Hill's equation (3.7) under the assumption that the parameter ϵ can be considered small, since ϵ is generally the parameter with respect to which the periodic solutions of the original nonlinear problem are developed.   In the present section we carry out such a discussion for the special case of the Mathieu equation and will then apply the results in the next section to the discussion of the stability of the harmonic solutions of the Duffing equation (cf. Chapter IV, Section 11).

Our object is to obtain the equations of the boundary curves of the stable regions of the Mathieu equation

$$(5.1) \qquad \frac{d^2 w}{dz^2} + (\delta + \epsilon \cos z)w = 0$$

in the form $\delta = \delta(\epsilon)$, and, since we assume ϵ to be small, we shall develop $\delta(\epsilon)$ in powers of ϵ and retain only a certain number of the terms of lowest order in the development. The discussion of the preceding section has furnished us with the following facts which we shall use to obtain the development of $\delta(\epsilon)$ in powers of ϵ: The stable regions are connected together at the points $\delta = \dfrac{n^2}{4}$, $\epsilon = 0$ ($n = 0$, 1, 2, $\cdots$) which obviously correspond to the linearly independent periodic solutions $\sin \dfrac{n}{2} z$, $\cos \dfrac{n}{2} z$ of (5.1). From each of the points $\delta = \dfrac{n^2}{4}$, $\epsilon = 0$ we expect two branches of the curves $\delta = \delta(\epsilon)$ to emerge except in the case $n = 0$ when only one branch is to be expected. Each point of the boundary curves $\delta = \delta(\epsilon)$ corresponds to a periodic solution of (5.1) of period 2π or of period 4π and, as we have seen near the close of the preceding section, these solutions are the unique periodic solutions (within a constant multiplier, that is) associated with such points for $\epsilon \neq 0$. As $\epsilon \to 0$, we expect that these periodic solutions will tend to appropriate multiples of either $\sin \dfrac{n}{2} z$ or $\cos \dfrac{n}{2} z$.

We assume that the solutions $w(z; \epsilon)$ of (5.1) as well as $\delta(\epsilon)$ can be expanded in series of powers of ϵ as follows:*

$$(5.2) \qquad \begin{cases} w = w_0 + \epsilon w_1 + \epsilon^2 w_2 + \cdots, \\ \delta = \delta_0 + \epsilon \delta_1 + \epsilon^2 \delta_2 + \cdots. \end{cases}$$

The quantities δ_i are constants and the quantities w_i are functions of z which must be determined in such a way that w is a solution of

---

* The proof in Appendix I for the existence of periodic solutions analytic in ϵ of differential equations of the form $\ddot{x} + x = \epsilon f(x, \dot{x}, t)$, with f periodic in t, holds for equation (5.1).

(5.1) with period 2π or 4π which reduces to $\cos \frac{n}{2} z$ or $\sin \frac{n}{2} z$ when $\epsilon \to 0$. Insertion of (5.2) in (5.1) yields the following relation:

$$(w_0'' + \epsilon w_1'' + \cdots) + [(\delta_0 + \epsilon \delta_1 + \cdots) + \epsilon \cos z]$$
$$\cdot (w_0 + \epsilon w_1 + \cdots) = 0$$

which is satisfied only if the coefficients of all powers of ϵ vanish; we are therefore led to the following differential equations for the functions w_i :

$$(5.3) \quad \begin{cases} w_0'' + \delta_0 w_0 = 0, \\ w_1'' + \delta_0 w_1 = -\delta_1 w_0 - w_0 \cos z, \\ w_2'' + \delta_0 w_2 = -\delta_2 w_0 - \delta_1 w_1 - w_1 \cos z, \\ \cdots\cdots\cdots\cdots\cdots\cdots\cdots , \\ \cdots\cdots\cdots\cdots\cdots\cdots\cdots . \end{cases}$$

In addition, we require that each function w_i should have the period 2π or 4π. The first equation of (5.3) leads therefore of necessity to the following values for δ_0 :

$$(5.4) \qquad \delta_0 = \frac{n^2}{4}, \qquad n = 0, 1, 2, \cdots ,$$

and to the following functions for w_0 :

$$(5.5) \quad \begin{cases} w_0 = \cos \frac{n}{2} z \\ w_0 = \sin \frac{n}{2} z \end{cases} \qquad n = 0, 1, 2, \cdots ,$$

as was to be expected. We observe that the solution w given by (5.2) reduces as it should to $\cos \frac{n}{2} z$ or $\sin \frac{n}{2} z$ when $\epsilon \to 0$. We proceed to the higher approximations for the cases $n = 0, 1, 2$:

$n = 0$: In this case $\delta_0 = 0$, $w_0 = 1$, and the equation for w_1 is

$$w_1'' = -\delta_1 - \cos z.$$

In order that w_1 be periodic it is clearly necessary that δ_1 be zero, in which case $w_1 = \cos z + c$, c a constant. The equation for w_2 becomes

$$w_2'' = -\delta_2 - (\cos z + c) \cos z$$

$$= -\delta_2 - \tfrac{1}{2} - c \cos z - \tfrac{1}{2} \cos 2z$$

from which we conclude that $\delta_2 = -\tfrac{1}{2}$ in order that w_2 should be periodic. Up to terms of second order in ϵ we have therefore

(5.6) $$\delta = -\tfrac{1}{2}\,\epsilon^2 + \cdots .$$

$n = 1$: In this case $\delta_0 = \tfrac{1}{4}$ and $w_0 = \cos\dfrac{z}{2}$ or $w_0 = \sin\dfrac{z}{2}$. If we take $w_0 = \cos\dfrac{z}{2}$ we find for w_1 the equation

$$w_1'' + \tfrac{1}{4}\,w_1 = (-\,\delta_1 - \cos z) \cos\frac{z}{2}$$

$$= (-\,\delta_1 - \tfrac{1}{2}) \cos\frac{z}{2} - \tfrac{1}{2} \cos\frac{3z}{2},$$

since $\cos\dfrac{z}{2} \cos z = \tfrac{1}{2}\left(\cos\dfrac{z}{2} + \cos\dfrac{3z}{2}\right)$. If the coefficient of $\cos\dfrac{z}{2}$ in the right-hand side of this equation were not zero the function w_1 would contain a term of the form $z \sin\dfrac{z}{2}$ corresponding to the fact that the homogeneous differential equation is satisfied by $\cos\dfrac{z}{2}$. To insure periodicity of w_1 it is therefore necessary to require that $\delta_1 + \tfrac{1}{2} = 0$. In this case we obtain for δ the relation

(5.7) $$\delta = \tfrac{1}{4} - \tfrac{1}{2}\epsilon$$

valid within terms of order ϵ^2 or higher. If we take $w_0 = \sin\dfrac{z}{2}$ rather than $w_0 = \cos\dfrac{z}{2}$ we find in the same way the relation

(5.8) $$\delta = \tfrac{1}{4} + \tfrac{1}{2}\epsilon$$

also valid within terms of order ϵ^2.

$n = 2$: In this case $\delta_0 = 1$ and $w_0 = \cos z$ or $w_0 = \sin z$. If we choose $w_0 = \cos z$ we would find by proceeding in exactly the same way as above that $\delta_1 = 0$, $\delta_2 = \dfrac{5}{12}\epsilon^2$, so that

(5.9) $$\delta = 1 + \frac{5}{12}\epsilon^2$$

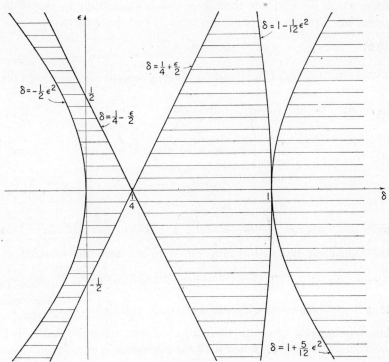

Fɪɢ. 5.1. Approximate regions of stability of the Mathieu equation for small ϵ.

within terms of order ϵ^3. On the other hand, the choice of $w_0 = \sin z$ z would lead to $\delta_1 = 0$, $\delta_2 = -\dfrac{1}{12}\epsilon^2$, so that

(5.10) $$\delta = 1 - \frac{1}{12}\epsilon^2$$

within terms of order ϵ^3.

The result of plotting the curves outlining the stable regions is indicated in Figure 5.1. The shaded regions are the stable regions in accordance with the theorems of Haupt discussed in Section 3) since they are the regions whose boundary points are correlated with solutions of (5.1) of different periods. The agreement with Figure 4.1 is quite good for not too large values of ϵ.

6. Stability of the harmonic solutions of the Duffing equation

In Chapter IV, Section 11, we have seen that the discussion of the stability of the harmonic solutions of the Duffing equation depends on the following Mathieu equation:

$$(6.1) \qquad \delta \ddot{x} + \left[\left(\alpha + \frac{3}{2} \beta A^2 \right) + \frac{3}{2} \beta A^2 \cos 2\omega t \right] \delta x = 0$$

for the variation $\delta x(t)$ of the harmonic solution $x(t)$, which was presumed to be given with sufficient accuracy by its approximation of lowest order $x = A \cos \omega t$. It was also found that the "amplitude" A of $x(t)$ depends upon the frequency ω and on the amplitude $F_0 \beta$ of the external periodic force $F_0 \beta \cos \omega t$ in accordance with the relation

$$(6.2) \qquad \omega^2 = \alpha + \tfrac{3}{4} \beta A^2 - F_0 \beta / A$$

under the assumption that the parameter β is so small that terms of order β^2 or higher could be ignored. For stability we require that all solutions δx of (6.1) remain bounded for $t > 0$. As we know from the preceding Section 5), this requires that the coefficients in (6.1) satisfy certain conditions which are readily obtained approximately if β is small enough. Our object is to decide which pairs of values of A and ω satisfying (6.2) lead to stable solutions of (6.1) and which to unstable solutions, or in other words to decide which portions of the response curves given by (6.2) correspond to stable and which to unstable regions. In Figure 6.1 we indicate schematically the character of the response curves given by (6.2) (cf. Chapter IV, Sect. 2) for a complete discussion of these curves). The curve for $F_0 \beta = 0$ indicates the response curve for the free oscillation when no external force acts on the system; the other curves indicate the two branches which occur for $\beta > 0$, one of them corresponding to an oscillation in phase with the external force $(A > 0)$, the other to an oscillation 180° out of phase with the external force $(A < 0)$. The points marked T are the points at which the response curves have a vertical tangent; they are characterized by $d\omega/dA = 0$, from which we obtain the relation

$$(6.3) \qquad \frac{3}{2}\beta A + F_0\beta/A^2 = 0$$

in view of (6.2). In Chapter IV we have already argued on physical grounds that the points T should be transition points from stable to unstable pairs of values of A and ω. This conjecture will be found to be correct in the course of the discussion to follow. We recall that the amplitude $F_0\beta$ of the excitation is always assumed to be positive.

In order to make direct use of the stability relations developed in the preceding Section 5) it is convenient to introduce a new inde-

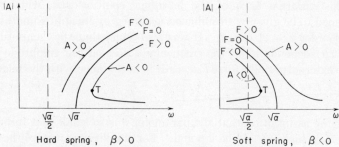

Fig. 6.1. Schematic response curves for the Duffing equation.

pendent variable replacing t and new quantities replacing the parameters occurring in (6.1), as follows:

$$(6.4) \qquad \begin{cases} z = 2\omega t, \\[2mm] 4\omega^2\delta = \alpha + \dfrac{3}{2}\beta A^2, \\[2mm] 4\omega^2\epsilon = \dfrac{3}{2}\beta A^2. \end{cases}$$

In terms of these quantities (6.1) becomes

$$(6.5) \qquad \frac{d^2\,\delta x}{dz^2} + (\delta + \epsilon \cos z)\,\delta x = 0,$$

as one can readily verify. We observe that ϵ is a small quantity proportional to β.   In the δ, ϵ-plane the set of curves corresponding to the response equation (6.2) is given in parametric form by

$$(6.6) \qquad \begin{cases} 4\delta = \dfrac{\alpha + \dfrac{3}{2}\beta A^2}{\alpha + \dfrac{3}{4}\beta A^2 - F_0\beta/A} \\[4ex] 4\epsilon = \dfrac{\dfrac{3}{2}\beta A^2}{\alpha + \dfrac{3}{4}\beta A^2 - F_0\beta/A} \end{cases}$$

with A as parameter, in view of (6.4).   Each different value of F yields of course a different curve.   In any given case the decision as to stability or instability depends upon whether the points $(\delta,\ \epsilon)$ given by (6.6) fall into the stable or the unstable regions of Figure 5.1.

We therefore turn to a geometrical discussion of the curves (6.6) in the δ, ϵ-plane.   The two cases of hard ($\beta > 0$) and soft ($\beta < 0$) springs are best discussed separately.   We begin with the case $\beta > 0$.   As A approaches zero through negative values it is readily seen that δ approaches zero through positive values and that all of the curves are tangent to the δ-axis at the origin.   As A varies in a monotone way from zero to $+\infty$ or from zero to $-\infty$, one sees readily that both δ and ϵ approach the point $(\frac{1}{2}, \frac{1}{2})$ independent of the value of $F_0\beta$.   The point $(\frac{1}{2}, \frac{1}{2})$ is singular, that is, $d\delta/dA$ and $d\epsilon/dA$ both vanish; the point is in fact easily seen to be a cusp at which two branches of the curves given by (6.6) come together with a common tangent having the slope $+2$; one of these is traversed as $A \to -\infty$, and the other as $A \to +\infty$.   As A decreases from $+\infty$, the values of δ and ϵ decrease at first, but later increase again.   In Figure 6.2 two curves of the set (6.6) have been plotted; in both cases the quantity $\kappa = 3\beta/2\alpha$ has been assumed to have the value 0.1, while two values, i.e. $\kappa_1 = 0.1$ and $\kappa_1 = 1.0$, were chosen for the quantity $\kappa_1 = F_0\beta/\alpha$.   The two curves thus represent response curves for two rather widely separated values for the amplitude of the external force.

We observe that for $A < 0$ the curve points lie to the left of the straight line $\delta = \frac{1}{4} + \epsilon/2$ and to the right of it for $A > 0$, and this behavior is typical for all of the response curves when $\beta > 0$.   We observe that the points on the response curves for $A > 0$ are stable (cf. Figure 6.1 also), at least for values of A which are not so small that the curve points come into the vicinity of $\delta = 1$. The points on the branches corresponding to $A < 0$ are seen to be stable until A reaches a value corresponding to the points marked T in Figure 6.2 where the response curve crosses the line $\delta = \frac{1}{4} - \epsilon/2$, after which the points are all unstable points. We proceed to show that *the points on $\delta = \frac{1}{4} - \epsilon/2$ correspond to the previously mentioned points with vertical tangents on the response curves shown in Figure* 6.1, i.e. to the transition from stable to unstable pairs of values of A and ω. For this purpose we develop the relations (6.6) with respect to β and neglect terms of order higher than the first (this is of course consistent with our general

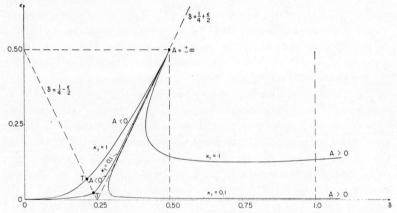

FIG. 6.2. Map of response curves for the Duffing equation $(\beta > 0)$ on the stability regions of the Mathieu equation.

procedure in which we have systematically ignored terms of this order in β); the result is

$$(6.7) \quad \begin{cases} 4\delta = 1 + \left(\dfrac{3}{4}\dfrac{A^2}{\alpha} + \dfrac{F_0}{A\alpha} \right)\beta, \\[2ex] 4\epsilon = \dfrac{3}{2}\dfrac{A^2}{\alpha}\beta, \end{cases}$$

as one can readily verify. We are at present interested in those values of A which satisfy (6.3), i.e. for which $3/2\, A^2 = -F_0/A$. In this case, one sees immediately that the curve points furnished by (6.7) satisfy the equation $\delta = \frac{1}{4} - \epsilon/2$, which proves our statement.

*The locus of the vertical tangents of the response curves in the A, ω-plane
thus maps on the boundary between a stable and an unstable region of the
δ, ε-plane,* and this verifies the conjectures in Chapter IV (cf. particularly Section 3 and Section 11).

We turn now to the discussion of stability for the case $\beta < 0$, i.e.
the case of a soft spring.   The curves corresponding to (6.6) in this

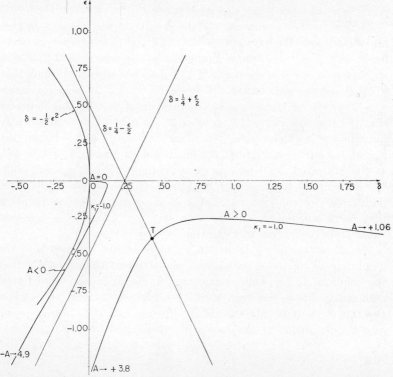

FIG. 6.3. Map of response curves for the Duffing equation $(\beta < 0)$ on the
stability regions of the Mathieu equation.

case are different from those in the preceding case because of the fact
that the denominators in the right-hand sides of (6.6) tend to zero
for three real values of A so that the curve points in the δ, ϵ-plane
tend to infinity when A approaches these values.   In addition, ϵ is
always negative.   In Figure 6.3 the curves have been plotted for the
case $\kappa = 3\beta/2\alpha = -0.1$ and $\kappa_1 = 1.0$.   Unlike the preceding case

for $\beta > 0$, the branch of the response curve with $A < 0$ (the "out of phase" branch) is stable, while the branch with $A > 0$ is stable only for amplitudes less (numerically) than the transition value T where the response curve has a vertical tangent. It has already been shown that the locus of the transition points T (when F_0 varies) falls on the line $\delta = \frac{1}{4} - \epsilon/2$ which separates a stable from an unstable region. Thus the results for stability in the case $\beta < 0$ are also shown to occur in accordance with our expectations.

A few further comments on the results of the present section are of interest. We have noted that the image of the response curves for $A > 0$ (i.e. the in-phase branches) in the case $\beta > 0$ falls in a stable region of the δ, ϵ-plane only if A does not become so small that δ lies near the value unity, which means that ω lies near the value $\sqrt{\alpha}/2$, as we see from the second equation of (6.4). This indicates that the response curves given by (6.2) for $\beta > 0$ are accurate only for values of ω which exceed somewhat the value $\sqrt{\alpha}/2$. In the case $\beta < 0$, the curves of Figure 6.3 indicate that a branch $A < 0$ as well as the branch $A > 0$ may become unstable when δ is near unity. One is tempted to conjecture that the response curves might show peculiarities of one kind or another in the vicinity of these points if they were calculated with greater accuracy. This is a question which deserves further investigation.

In all of the above discussion we have tacitly assumed that the amplitude βF_0 of the external force was different from zero, i.e. that the solutions of the Duffing equation under discussion were forced oscillations. It is, however, also of considerable interest to consider the case $F_0 = 0$, i.e. the case of the free oscillations. In this special case the parameter A can be readily eliminated from (6.6) to yield

$$(6.8) \qquad\qquad \delta = \frac{1}{4} + \frac{\epsilon}{2}.$$

A comparison with Figure 6.2 shows that these points lie on the boundary between stable and unstable regions and hence as we know are points corresponding to unstable solutions. We have thus obtained the rather paradoxical result that the free oscillations of the Duffing equation are unstable! Of course the reason for this lies in our definition of stability. We propose to discuss this point in detail in the next section, where we introduce a modified but still quite reasonable definition for stability which leads to exactly the same results for stability of the forced oscillations of the Duffing equation as we have found in this section, while the free oscillations are found to be stable on the basis of the new definition.

7. Orbital stability of the harmonic solutions of the Duffing equation

In the preceding section we have seen that the free oscillations of the Duffing equation are unstable on the basis of the definition of stability used there. This seems rather strange at first sight since all of these solutions are periodic and they therefore can be represented in the velocity-displacement plane as a set of closed curves; a slight disturbance therefore has the effect of a shift from one such closed curve in the phase plane to another which lies in its neighborhood. However, the periods of the two oscillations differ slightly, as we know, and thus one sees readily that the corresponding displacements will differ by finite amounts for sufficiently large values of the time even though the disturbance is made very small. It is thus clear that the definition of stability used hitherto must yield the result that such oscillations are unstable, since the motion to be tested for stability is always compared with another *for the same values of the time*. On the other hand, it would seem not unreasonable in many cases to consider a given motion as stable if the perturbed motions yield curves *in the phase plane* which lie near to the image of the original curve in that plane—in other words if the given motion and all neighboring ones are such that the velocities and displacements differ only slightly throughout the motion when the time variable is appropriately transformed in each particular neighboring solution. In astronomy the problem of the stability of the motion of the planets is one in which such a definition of stability is clearly appropriate, and in fact the stability problem is often referred to in these cases as a problem in *orbital stability*.

We turn then to the discussion of the orbital stability of the harmonic solutions of the Duffing equation:*

$$(7.1) \qquad \ddot{x} + \alpha x + \beta x^3 = F \cos \omega t.$$

Up to now we have considered a harmonic solution $x(t)$ of (7.1)

---

* In Appendix V we derive the Poincaré criterion for orbital stability in all generality and without specific reference to the theory of Hill's equation.

and compared it with a neighboring solution $x(t) + \delta x(t)$ obtained by changing the initial data slightly; the solution $x(t)$ was then said to be stable or unstable according to whether the solutions of the variational equation, in this case a Hill's equation:

$$(7.2) \qquad \delta\ddot{x} + (\alpha + 3\beta x^2)\delta x = 0$$

were all bounded or not. For the discussion of orbital stability it is necessary, in accordance with our above discussion, to consider variations of a more general character than has been customary, in which the independent variable (the time) as well as the initial conditions are varied. We consider therefore a set of solutions of (7.1) of the form $x(t; \lambda_1, \lambda_2)$ in which λ_1 and λ_2 depend on the initial values of x and $\dot{x}$ for $t = 0$ in such a way that the harmonic solution $x_0(t)$ whose stability is in question is obtained for $\lambda_1 = \lambda_2 = 0$, i.e. $x_0(t) = x(t; 0, 0)$. We then consider variations of these solutions of the form $x(f(\lambda_1, \lambda_2)t; \lambda_1, \lambda_2)^*$ in which the factor $f(\lambda_1, \lambda_2)$ on t is a function which reduces to unity when λ_1 and λ_2 are zero; our object then is to decide whether a function $f(\lambda_1, \lambda_2)$ can be found for sufficiently small but otherwise arbitrary values of λ_1 and λ_2 such that $|x - x_0|$ remains bounded for all t, at least within terms of first order in the small quantities λ_1 and λ_2. Within first order terms in λ_1 and λ_2 we may write

$$(7.3) \qquad f(\lambda_1, \lambda_2) = 1 + \kappa_1\lambda_1 + \kappa_2\lambda_2$$

with κ_1 and κ_2 certain constants, and hence we have also with the same degree of accuracy the relation

$$
\begin{aligned}
x(f(\lambda_1, \lambda_2)t; \lambda_1, \lambda_2) &- x_0(t) \\
(7.4) \qquad &= \left(t\kappa_1\dot{x}_0 + \frac{\partial x_0}{\partial\lambda_1}\right)\lambda_1 + \left(t\kappa_2\dot{x}_0 + \frac{\partial x_0}{\partial\lambda_2}\right)\lambda_2
\end{aligned}
$$

in which of course $\dot{x}_0, \dfrac{\partial x_0}{\partial\lambda_1}, \dfrac{\partial x_0}{\partial\lambda_2}$ are all evaluated for $\lambda_1 = \lambda_2 = 0$.

Our problem in any given case is to decide whether the two terms on the right-hand side of (7.4) can be made bounded or not for all t by making appropriate choices for the constants κ_1 and κ_2; if they can be made bounded, we say that the solution is orbitally stable.

---

* These functions will in general not be solutions of (7.1), of course.

We begin by observing that $\delta x = \dfrac{\partial x_0}{\partial \lambda_1}$ and $\delta x = \dfrac{\partial x_0}{\partial \lambda_2}$ are certain special solutions of (7.2) when $x = x_0(t)$, as one can readily verify simply by differentiating (7.1) with respect to λ_1 and λ_2 and observing that the right-hand side of (7.1) is independent of λ_1 and λ_2. Consider now the case of the free oscillations of the Duffing equation, i.e. the case in which $F = 0$. In this case we observe that $\dot{x}_0$ is also a solution of (7.2), as we can see by differentiating (7.1) with respect to t; thus the parameters α and β in (7.2) have such values that (7.2) possesses a periodic solution having the same period as the coefficient $\alpha + 3\beta x^2$, since $\dot{x}_0$ has the same period as x_0. It follows (cf. (3.5) and the subsequent discussion) that the characteristic equation for (7.2) has a double root and hence (cf. (2.14)) fundamental solutions δx_1, δx_2 of (7.2) exist which have the form

(7.5)
$$\begin{cases} \delta x_1 = \dot{x}_0(t), \\ \delta x_2 = At\dot{x}_0(t) + \psi(t), \end{cases}$$

in which $\psi(t)$ is a function having the same period as $x_0(t)$ and A is a certain constant which may or may not be zero. The fixed functions $\dfrac{\partial x_0}{\partial \lambda_1}$ and $\dfrac{\partial x_0}{\partial \lambda_2}$ are solutions of (7.2), as we have already remarked, and hence are certain linear combinations of δx_1 and δx_2. From the form of the quantities in the two parentheses on the right-hand side of (7.4) we now see that κ_1 and κ_2 can always be chosen in such a way that the only term which could possibly be unbounded—i.e. a term of the form $Bt\dot{x}_0(t)$—will be cancelled out. It follows therefore that the free oscillations of the Duffing equation are stable on the basis of the new definition, i.e. they are orbitally stable.

It is obvious that any solutions of (7.1) which were stable in the earlier sense will also be orbitally stable: one need only choose κ_1 and κ_2 to be zero in these cases. It is also true that the new definition of stability does not cause solutions of (7.1) other than the free oscillations to become stable which were unstable according to the earlier definition: this follows at once from the known fact that any unbounded solutions $\dfrac{\partial x_0}{\partial \lambda_i}$ of (7.2) which are correlated with points in the *interior* of an unstable region of the Mathieu equation become exponentially infinite in these cases, so that it is not possible to make the quantities $t\kappa_i\dot{x}_0 + \dfrac{\partial x_0}{\partial\lambda_i}$ in (7.4) bounded by any choice of κ_i in view of the fact that $\dot{x}_0$ is a periodic function. It follows therefore that the two definitions of stability yield the same results for the harmonic solutions of the Duffing equation except in the case of the free oscillations, which are stable rather than unstable on the basis of the less restrictive definition.

APPENDIX I

Mathematical Justification of the Perturbation Method

1. Existence of the perturbation series in general

In this book we have in general assumed without proof that the methods of solution involving infinite processes of one sort or another, e.g., the iteration method or the perturbation method, converge to the desired solution. In this section we show how one of them, the perturbation method, can be given a rigorous mathematical justification. In the next section we shall see that important information regarding the mode of attack that is appropriate in various special cases can also be obtained from some of the formulas used in the existence proof.*

We consider the differential equation

$$(1.1) \qquad \ddot{x} + x = \epsilon f(x, \dot{x}, \theta), \qquad \theta = \omega t$$

for $x(t)$, in which the function f is assumed to have the period $2\pi/\omega$ in t and ϵ is a small parameter. This differential equation is clearly general enough to include all of the cases treated in this book whenever a development in the neighborhood of the linearized vibration problem was in question. It also includes the case of the Hill's equation for small periodic coefficients.

Our aim is to prove the existence of solutions of period $2\pi/\omega$ expressible as power series in ϵ. Instead of proving directly the convergence of such power series, it is simpler to prove that there exist periodic solutions which depend analytically upon ϵ, which then implies the possibility of a power series expansion. It is therefore appropriate to assume that $f(x, \dot{x}, \theta)$ is an analytic function of the three complex variables $x, \dot{x}$, and θ for all values of these arguments; the parameter ϵ is also to be considered complex. It follows** that all solutions $x(t; \epsilon)$ of (1.1) depend analytically upon t and ϵ.

* The basic ideas in this proof are due to Poincaré [33]. The presentation given here follows that given by Friedrichs [40].
**For a proof of this, see Ince [17], Ch. XII.

223

We are interested in establishing the existence of periodic solutions $x(t)$. It follows that we may assume the initial condition $\dot{x}(0) = 0$ without loss of generality, provided that we shift the phase of f in (1.1) appropriately. We write, therefore, in place of (1.1)

$$(1.2) \qquad\qquad \ddot{x} + x = \epsilon f(x, \dot{x}, \theta + \delta),$$

in which the phase shift δ is to be chosen properly. Periodic solutions of (1.2) are to be found which satisfy the initial conditions

$$(1.3) \qquad\qquad x(0) = A, \qquad \dot{x}(0) = 0,$$

the complex number A being at our disposal. It is known (see, for example, the book of Ince cited above) that a unique solution $x(t) = x(t; A, \delta, \epsilon)$ of (1.2) satisfying the initial conditions (1.3) exists in the domain

$$|t| \le T_1, \qquad |x - A| \le A_1, \qquad |\dot{x}| \le A_1, \qquad A_1 = 3\,|A|,$$

in which $x(t; A, \delta, \epsilon)$ depends analytically on t, A, δ, and ϵ. For $\epsilon = 0$, in particular, the solution is

$$(1.4) \qquad\qquad x(t) = A \cos t,$$

and it has the real period 2π. Our purpose is to establish the existence of periodic solutions in a neighborhood of $\epsilon = 0$ by assuming that $A = A(\epsilon)$, $\delta = \delta(\epsilon)$, and selecting from the solutions $x(t; A(\epsilon), \delta(\epsilon), \epsilon)$ of the initial value problem those which are periodic. The period T will then also be a function of ϵ: $T = T(\epsilon) = 2\pi/\omega$, which must approach 2π as ϵ approaches zero. It must therefore be shown that the bound T_1 for $|t|$ can be chosen greater than 2π. We indicate how this can be done by taking ϵ sufficiently small, on the basis of two relations which are of fundamental importance otherwise for the entire discussion to follow. The differential equation (1.2) and initial conditions (1.3) are replaced by the equivalent integral equation

$$(1.5) \qquad x(t) = A \cos t + \epsilon \int_0^t f[x(\tau), \dot{x}(\tau), \omega\tau + \delta] \sin(t - \tau)\, d\tau,$$

from which

$$(1.6) \qquad \dot{x}(t) = -A \sin t + \epsilon \int_0^t f[x(\tau), \dot{x}(\tau), \omega\tau + \delta] \cos(t - \tau)\, d\tau.$$

If ϵ is chosen small enough, e.g., if $\epsilon \le A/3\pi M$, M being the maximum of $f(x, \dot{x}, \theta)$ in the domain $|x - A| \leq A_1, |\dot{x}| \leq A_1, -\infty < \theta < \infty$, it is easily seen that iterations applied to (1.5) in the usual way will lead to a sequence of functions $x_n(t)$ with values which stay within the domain $|x - A| \leq A_1, |\dot{x}| \leq A_1$, if $|t| \leq 3\pi$. It then follows along standard lines that the solution $x(t)$ exists for $|t| \leq 3\pi$.

We now seek functions $A(\epsilon), \delta(\epsilon), T(\epsilon)$ such that $x(t; A(\epsilon), \delta(\epsilon), \epsilon)$ is periodic with the period $T = 2\pi/\omega$. The conditions of periodicity are

$$(1.7) \qquad \begin{cases} x(T; A, \delta, \epsilon) = x(0; A, \delta, \epsilon), \\ \dot{x}(T; A, \delta, \epsilon) = 0. \end{cases}$$

Our problem would clearly be solved if we could show that analytic functions $A(\epsilon)$, $T(\epsilon)$, and $\delta(\epsilon)$ can be chosen in such a way as to satisfy equations (1.7). These periodicity relations are two equations for the three functions A, δ, and T, so that there is some latitude possible in deciding what quantity should be prescribed in advance. For instance, one might prescribe A arbitrarily and seek to determine $T(\epsilon)$ and $\delta(\epsilon)$, as was done actually in Chapter IV in dealing with forced oscillations of systems with nonlinear restoring forces; or one might prescribe δ and find $A(\epsilon)$ and $T(\epsilon)$ as was done in Chapter V in dealing with the free oscillations of self-sustained systems. In Chapter IV we found also that it was not a matter of indifference whether one choice or another for the quantity to be prescribed was made—for the cases considered there it was in fact found really essential to prescribe the quantity A and then determine T in accord with that choice. As we have already hinted above, the method we follow here to establish the validity of the perturbation series also yields clues regarding the proper procedure to be used in individual cases in order to obtain the series concretely.

We have reduced the problem of determining the existence of periodic solutions of our differential equation to the problem of solving the periodicity equations (1.7) in a neighborhood of $\epsilon = 0$. To solve the equations we naturally wish to make use of the implicit function theorem in a neighborhood of $\epsilon = 0$. However, this theorem cannot be applied to equations (1.7) in their present form since they are, as we know, satisfied identically in A for $T = 2\pi$ (i.e., in the case $\epsilon = 0$). This difficulty can be overcome by proceeding as follows. Instead of T a new variable η is introduced by the equation

$$(1.8) \qquad T = 2\pi + \epsilon\eta(\epsilon),$$

so that the frequency ω is given by

$$(1.8)' \qquad \omega = \frac{1}{1 + \dfrac{\epsilon\eta}{2\pi}} \simeq 1 - \frac{\epsilon\eta}{2\pi} \qquad \text{for } \epsilon \text{ small.}$$

Thus $T = 2\pi$ automatically when $\epsilon = 0$. In order to formulate periodicity conditions in such a way that they are not satisfied identically we introduce the quantities

$$(1.9) \qquad P = \epsilon^{-1}[A - x(T; A, \delta, \epsilon)] = P(A, \delta, \eta; \epsilon),$$

$$(1.10) \qquad Q = -\epsilon^{-1}[\dot{x}(T; A, \delta, \epsilon)] = Q(A, \delta, \eta; \epsilon),$$

so that the periodicity conditions are now given by

$$(1.11) \qquad P = Q = 0.$$

The choice made for P and Q was determined by the integral representations (1.5) and (1.6) for $x(t)$ and $\dot{x}(t)$: we shall see in a moment that the definitions of P and Q can indeed be extended so that these functions are regular and analytic for $\epsilon = 0$ and that they do not in general become identities in A for $\epsilon = 0$. The quantities P and Q are in fact given by

$$(1.12) \qquad \begin{aligned} P = {}& A\epsilon^{-1}[(1 - \cos \epsilon\eta)] \\ & + \int_0^{2\pi+\epsilon\eta} f[x(\tau), \dot{x}(\tau), \omega\tau + \delta] \sin (\tau - \epsilon\eta) \, d\tau, \end{aligned}$$

$$(1.13) \qquad \begin{aligned} Q = {}& A\epsilon^{-1} \sin \epsilon\eta \\ & - \int_0^{2\pi+\epsilon\eta} f[x(\tau), \dot{x}(\tau), \omega\tau + \delta] \cos (\tau - \epsilon\eta) \, d\tau, \end{aligned}$$

with ω expressed in terms of η through (1.8)'. Since $\epsilon^{-1} \sin \epsilon$ and $\epsilon^{-1}(1 - \cos \epsilon)$ can be extended to $\epsilon = 0$ as regular analytic functions of ϵ, the same is true for P and Q considered as functions of the complex variables ϵ, η, and δ.

With P and Q defined by (1.12) and (1.13) we can in principle investigate the periodicity conditions $P = Q = 0$, beginning with the equations for $\epsilon = 0$. If it could be shown that the equations for $\epsilon = 0$ possess a solution for any one of the possible pairs of values (A, η), (A, δ), (δ, η)—and for arbitrary values of the respective third quantity—such that the Jacobian J_0 of P and Q with respect to the appropriate pair of values does not vanish for $\epsilon = 0$, it would follow from the implicit function theorem that the periodicity equations could be solved in a neighborhood of $\epsilon = 0$ to yield $(A(\epsilon),\ \eta(\epsilon))$, $(A(\epsilon),\ \delta(\epsilon))$, or $(\delta(\epsilon),\ \eta(\epsilon))$. Under these circumstances the existence of periodic solutions of (1.1) would be established; these solutions would have the period $T = 2\pi + \epsilon\eta$ and they would be regular analytic functions of ϵ in a neighborhood of $\epsilon = 0$ and therefore would possess power series developments in ϵ. In any given case, then— i.e. for any given f—the existence of the perturbation series may be decided by studying a certain Jacobian. Once the existence of the series is established the successive coefficients of the series can be obtained by the formal processes illustrated in Chapter IV and Chapter V.

Before passing on to the study of concrete cases in the next section it is of interest to comment on the possibility of determining the periodic solutions in terms of quantities other than A, T, or δ. For instance, in Chapter IV we more often than not prescribed the coefficient A_1 of the term $A_1 \cos 2\pi t/T$ in the Fourier series for the periodic solution, instead of the quantity A. It would not be difficult to justify this procedure in these cases by showing that $A_1(\epsilon)$ is an analytic function of ϵ which reduces to A for $\epsilon = 0$ and that the function $A_1(A)$ can be inverted, i.e. that A can be considered a function of A_1. If this were done it is clear that A_1 could be prescribed instead of A and again the solution x and its period T would depend analytically on ϵ.

2. Existence of the perturbation series in concrete cases

Our object here is to study the solutions near $\epsilon = 0$ of the equations obtained from (1.12) and (1.13):

$$(2.1) \qquad P(A, \delta, \eta; \epsilon) = A\epsilon^{-1}(1 - \cos \epsilon\eta)$$

$$+ \int_0^{2\pi+\epsilon\eta} f[x(\tau), \dot{x}(\tau), \omega\tau + \delta] \sin (\tau - \epsilon\eta)\ d\tau = 0,$$

$$(2.2) \qquad Q(A, \delta, \eta; \epsilon) = A\epsilon^{-1} \sin \epsilon\eta$$

$$- \int_0^{2\pi+\epsilon\eta} f[x(\tau), \dot{x}(\tau), \omega\tau + \delta] \cos (\tau - \epsilon\eta)\ d\tau = 0,$$

in which $\omega = 2\pi/(2\pi + \epsilon\eta)$. We denote by P_0 and Q_0 the values of P and Q for $\epsilon = 0$; these quantities are thus given as follows:

$$(2.3) \quad P_0 = \int_0^{2\pi} f(A \cos \tau, -A \sin \tau, \tau + \delta) \sin \tau \, d\tau = P_0(A, \delta),$$

$$(2.4) \quad Q_0 = A\eta - \int_0^{2\pi} f(A \cos \tau, -A \sin \tau, \tau + \delta) \cos \tau \, d\tau$$
$$= Q_0(A, \delta, \eta),$$

since $x = A \cos t$, $\dot{x} = -A \sin t$ for $\epsilon = 0$. It is of some importance for the later discussion to observe that P_0 is independent of η, and that we may be able in some cases to solve the equation $P_0 = 0$ to obtain $A = A_0$ and insert its value in the equation $Q_0 = 0$ to determine η_0. If it is possible to do so, it is clear that the existence of a periodic solution is established and at the same time the lowest order terms A_0 and η_0 in the developments for the "amplitude" and for η (essentially the period) are obtained.

We proceed to discuss a number of special cases which arise from special choices of the function f. To begin with we consider these cases in two main groups: A) the free oscillations for which f is independent of t, and B) the forced oscillations in which f depends explicitly on t.

A. Free oscillations

In the case of the free oscillations, which are characterized by a function f not depending on θ, the quantities P_0 and Q_0 reduce to

$$(2.5) \qquad P_0(A) = \int_0^{2\pi} f(A \cos \tau, -A \sin \tau) \sin \tau \, d\tau,$$

$$(2.6) \quad Q_0(A, \eta) = A\eta - \int_0^{2\pi} f(A \cos \tau, -A \sin \tau) \cos \tau \, d\tau.$$

The simplest case to consider would seem to be the case in which f is also independent of $\dot{x}$, i.e. the case of the free oscillation without damping. In this case we have, however,

$$(2.7) \quad P_0(A) = \int_0^{2\pi} f(A \cos \tau) \sin \tau \, d\tau = -\int_{\tau=0}^{\tau=2\pi} f(A \cos \tau) \, d \cos \tau$$

and one sees that $P_0(A)$ is identically zero. Consequently our procedure fails; we shall see later how it can be modified in such a way as to yield the desired development in this case.

We turn next to cases in which damping occurs:

Case 1. Here we take the case of linearly damped oscillations, with $f \equiv f(x) - c\dot{x}$, $c =$ constant. One finds readily for P_0 the equation

$$(2.8) \qquad P_0(A) = cA\pi,$$

and since $P_0 = 0$ only for $A = A_0 = 0$ we obtain the expected result that no free linearly damped oscillation exists except the state of equilibrium.

Case 2. We suppose now that f represents a pure damping force, i.e. $f \equiv f(\dot{x})$, and for P_0 and Q_0 we have

$$(2.9) \qquad \begin{cases} P_0(A) = \displaystyle\int_0^{2\pi} f(-A \sin \tau) \sin \tau \, d\tau, \\[2mm] Q_0(A, \eta) = A\eta - \displaystyle\int_0^{2\pi} f(-A \sin \tau) \cos \tau \, d\tau. \end{cases}$$

If f represents a true damping force, i.e. a force which is always opposite in direction to the velocity, then $f(\dot{x})$ and $\dot{x}$ are opposite in sign and $P_0(A)$ would not vanish except for $A = A_0 = 0$. Again we have the result to be expected: no oscillation exists except the state of rest. However, if f represents a partly negative resistance, it may well be that the equation $P_0(A) = 0$ has a solution A_0 other than $A_0 = 0$. For example, in the case of the free oscillations of the van der Pol equation $f(\dot{x})$ is given by

$$(2.10) \qquad f(\dot{x}) \equiv (\dot{x} - \tfrac{1}{3} \dot{x}^3).$$

From (2.9) one finds for $P_0(A)$ the equation

$$(2.11) \qquad P_0(A) = -\pi A(1 - \tfrac{1}{4} A^2)$$

so that $P_0(A) = 0$ yields either $A_0 = 0$, or $A_0 = \pm 2$. With $A_0 = +2$ one then finds from $Q_0(2, \eta) = 0$ the value $\eta_0 = 0$ for η, which means that the frequency ω is given by $\omega = 1$ within terms of second order in ϵ. These results are identical with those obtained by the formal perturbation procedure in Part A of Chapter V.

Case 3. We now assume that f is an even function of $\dot{x}$, i.e. $f(x, -\dot{x}) = f(x, \dot{x})$. This includes the case mentioned above (i.e. f independent of $\dot{x}$) in which our procedure failed so far to give the desired development. A modified procedure is, however, easy to devise. We observe first that $x(-t)$ is a solution of (1.2) satisfying (1.3) if $x(t)$ is a solution and our condition on f holds. Since we are interested only in periodic solutions we may assume $\dot{x}(0) = 0$ without loss of generality. From the uniqueness theorem for the solution of the initial value problem of our differential equation it therefore follows that $x(-t) \equiv x(t)$, so that, in particular, $x(-T/2) = x(T/2)$ and $\dot{x}(-T/2) = -\dot{x}(T/2)$. On the other hand we have $\dot{x}(T + t) = \dot{x}(t)$, and consequently for $t = -T/2$ the relation $\dot{x}(-T/2) = \dot{x}(T/2)$. It follows therefore that $\dot{x}(T/2) = 0$, and the periodicity conditions are reduced in the present case to this one condition. The periodicity condition can therefore be written (cf. (1.13)) in the form

$$(2.12) \qquad R(A, \eta; \epsilon) = A\epsilon^{-1} \sin \frac{\epsilon\eta}{2}$$
$$- \int_0^{\pi+(\epsilon\eta/2)} f(x, \dot{x}) \cos\left(\tau - \frac{\epsilon\eta}{2}\right) d\tau = 0,$$

and this reduces for $\epsilon = 0$ to

$$(2.13) \quad R_0(A) = \frac{\eta}{2} - \int_0^\pi f(A \cos \tau, -A \sin \tau) \cos \tau \, d\tau = 0.$$

Now we are permitted to prescribe $A \neq 0$ arbitrarily, after which $R_0 = 0$ can be solved to yield

$$(2.14) \qquad \eta_0 = 2A^{-1} \int_0^\pi f(A \cos \tau, -A \sin \tau) \cos \tau \, d\tau.$$

Since $\dfrac{\partial R}{\partial \eta}\bigg|_{\epsilon=0} = \frac{1}{2} A \neq 0$, it follows that (2.12) can be solved for $\eta = \eta(\epsilon)$ with $\eta(0) = \eta_0$. This procedure would, in particular, yield the perturbation series for the free undamped oscillations of systems with nonlinear restoring forces with results coinciding with those obtained in Chapter IV.

B. Forced oscillations

In the case of forced oscillations, in which the function f is not independent of θ, we must solve the equations

$$(2.15) \quad P_0(A, \delta) = \int_0^{2\pi} f(A \cos \tau, -A \sin \tau, \tau + \delta) \sin \tau \, d\tau = 0,$$

$$(2.16) \quad \begin{aligned} Q_0(A, \delta, \eta) &= A\eta \\ &- \int_0^{2\pi} f(A \cos \tau, -A \sin \tau, \tau + \delta) \cos \tau \, d\tau = 0 \end{aligned}$$

obtained from (1.12) and (1.13) for $\epsilon = 0$. Again we consider a few special cases:

Case 1. Consider first the case of forced oscillations of a system with nonlinear damping but linear restoring force, i.e. the case in which $f \equiv f(\dot{x}) + F \cos (\omega t + \delta)$. The equation $P_0 = 0$ reduces to

$$(2.17) \quad \pi F \sin \delta = \int_0^{2\pi} f(-A \sin \tau) \sin \tau \, d\tau$$

and this equation has in general a solution A_0 even if f represents true damping, i.e. if $f(\dot{x})/\dot{x} < 0$. If f represents a partly negative resistance, as in the case of forced oscillations of self-excited systems, there may be several solutions if F is sufficiently small. For example, if f is given by

$$(2.18) \quad f \equiv \dot{x} - \tfrac{1}{3} \dot{x}^3 + F \cos (\omega t + \delta),$$

the equation $P_0 = 0$ for A_0 reduces to

$$(2.19) \quad A_0(1 - \tfrac{1}{4} A_0^2) = -F \sin \delta.$$

From (2.16) we find that

$$(2.20) \quad \eta_0 = FA_0^{-1}\pi \cos \delta.$$

The relations (2.19) and (2.20) yield the response curves for the forced oscillations, at least to the lowest order of approximation. In fact, we may set (cf. (1.18)') $2\pi\left(\dfrac{1 - \omega}{\epsilon}\right) \simeq \eta_0$ so that η_0 is proportional to the quantity σ of part B of Chapter V which was called the detuning, while A_0 can be interpreted, of course, as the amplitude of the response. If δ is then eliminated between (2.19) and (2.20) the result is readily found to be, with $\sigma = \eta_0/\pi$:

$$(2.21) \qquad A_0^2[\sigma^2 + (1 - \tfrac{1}{4} A_0^2)^2] = F^2$$

and this yields exactly the same response curves as were found in Part B of Chapter V when due regard is paid to differences in notation. This should not be surprising since the differential equation $\ddot{x} - (\dot{x} - \tfrac{1}{3} \dot{x}^3) + x = F \cos (\omega t + \delta)$ yields by differentiation the equation $\ddot{y} - (1 - y^2)\dot{y} + y = -F\omega \sin (\omega t + \delta)$ for $y = \dot{x}$ which is essentially the equation dealt with in Chapter V. The perturbation procedure thus furnishes the same result in the lowest order as the quite different procedure of van der Pol.

Case 2. Next we consider the case of the Duffing equation with linear damping, in which f is defined by

$$(2.22) \qquad f \equiv f(x) - c\dot{x} + F \cos (\theta + \delta), \qquad c \neq 0.$$

The equations $P_0 = 0$ and $Q_0 = 0$ are in this case

$$(2.23) \qquad P_0(A, \delta) = \pi c A - \pi F \sin \delta = 0,$$

$$(2.24) \quad Q_0(A, \delta, \eta) = A\eta - \int_0^{2\pi} f(A \cos \tau) \cos \tau \, d\tau - \pi F \cos \delta = 0,$$

as one finds with no difficulty. If $\delta = \delta_0 \neq 0$ is prescribed, (2.23) yields for A_0 the value

$$(2.25) \qquad A_0 = c^{-1} F \sin \delta_0, \qquad \text{with } A_0 \neq 0,$$

and this value inserted in (2.24) yields for η_0 the value

$$(2.26) \qquad \eta_0 = A_0^{-1}\left[\int_0^{2\pi} f(A_0 \cos \tau) \cos \tau \, d\tau + \pi F \cos \delta_0\right].$$

The last two equations yield the response curves to the lowest order in ϵ; one can easily verify that they give the same results as were found formally in Chapter IV for $c \neq 0$. If $c = 0$, we see from (2.23) that $\delta = 0$ or $\delta = \pi$, and (2.24) then yields the value of η_0 for values of $A \neq 0$ but otherwise arbitrary.

Case 3. Finally, we consider briefly the method to be followed in order to obtain subharmonic periodic solutions, for which a somewhat different approach is necessary. To begin with, the differential equation is written in the form

(2.27) $\qquad \ddot{x} + x - H \cos n\theta = \epsilon f(x, \dot{x}), \qquad \theta = \omega t,$

with n an integer larger than one, and H assumed to be kept fixed as $\epsilon \to 0$. Again we introduce a phase shift δ by replacing θ in (2.27) with $\theta + \delta$ and require

(2.28) $\qquad\qquad x(0) = A, \qquad \dot{x}(0) = 0.$

We introduce further the period $T = 2\pi/\omega$ of the oscillation desired and set

(2.29) $\qquad\qquad T = 2\pi + \epsilon\eta(\epsilon).$

For $\epsilon = 0$ the solution of (2.27) (with $\theta + \delta$ instead of θ) is

(2.30)
$$x(t) = x_0(t) = (A + kH \cos n\delta) \cos t$$
$$- nkH \sin n\delta \sin t - kH \cos n(t + \delta),$$

with $k = 1/(n^2 - 1)$.

The conditions under which a solution of (2.27) will have the period T, which is $1/n$ times the period of the excitation $H \cos n\omega t$, can then be formulated in the same way as above by two equations $P = 0, Q = 0$ which reduce for $\epsilon = 0$ to

(2.31)
$$\begin{cases} P_0(A, \delta, \eta) = \int_0^{2\pi} f[x_0(\tau), \dot{x}_0(\tau)] \sin \tau \, d\tau + \dfrac{n}{n^2 - 1} H \eta \sin \eta\delta \\[2ex] Q_0(A, \delta, \eta) = \left(A - \dfrac{H \cos n\delta}{n^2 - 1}\right)\eta \\[2ex] \qquad\qquad - \int_0^{2\pi} f[x_0(\tau), \dot{x}_0(\tau)] \cos \tau \, d\tau = 0. \end{cases}$$

The further discussion then follows the same lines as above.

APPENDIX II

The Existence of Combination Oscillations

In Appendix I the existence of periodic solutions of a nonlinear differential equation was proved for the case in which the excitation was a periodic function of the time. In this appendix we shall prove the existence of a special type of combination oscillations in the form of certain *almost periodic solutions* of the nonlinear differential equation

$$(1) \qquad \ddot{x} + c\dot{x} + x - \beta x^2 = h(t), \qquad c > 0,$$

with $h(t)$ an, in general, *almost periodic* function given by

$$(2) \qquad h(t) = \sum_{p_1, p_2} H_{p_1 p_2} e^{i(p_1 \omega_1 + p_2 \omega_2)t},$$

in which p_1, p_2 are certain positive or negative integers. At the end of Chapter IV in dealing with the problem of combination tones it was pointed out that the usual methods of approximation applied to equation (1) for $c = 0$ (i.e. with damping not present) and ω_1/ω_2 irrational would almost certainly lead to divergent series because of the occurrence of certain small divisors in the representations of the terms in the series expansions. The purpose of this appendix is to show that the "difficulty of the small divisors" can be overcome in some cases at least in a system which is provided with viscous damping.

To this end we follow a procedure given by Friedrichs [40]. First of all a set of functions $z(t)$ of the form

$$(3) \qquad z(t) = \sum_{p_1, p_2} C_{p_1 p_2} e^{i(p_1 \omega_1 + p_2 \omega_2)t} \equiv \{C_{p_1 p_2}\}$$

is introduced. For reasons which will soon be apparent it is useful to introduce a norm $\| z \|$ for the functions z by the definition

$$(4) \qquad \| z \| = \sum_{p_1, p_2} |C_{p_1 p_2}|,$$

and to denote by K the class of functions which have a finite norm. In particular, the external force, or excitation, $h(t)$, which is given in our notation by

$$(5) \qquad h(t) = \{H_{p_1 p_2}\},$$

is assumed to belong to the class K; we set

$$(6) \qquad \| h \| = \eta < \infty.$$

The class K of functions z is readily seen to be *closed* in the following sense: If $\{z_k\}$ is a set of functions belonging to K such that

$$\| z_n - z_m \| \to 0$$

when $m, n \to \infty$, then there exists a function z in K such that

$$\| z_n - z \| \to 0.$$

In other words, if the Cauchy criterion for convergence (in the sense of the norm defined by (4)) is satisfied, there exists a limit function of the class. Since, in addition,

$$| z(t) | \leq \| z \|$$

in view of (4), it follows that *convergence in the sense of the norm implies uniform convergence* in the present case.

The motivation for the choice of the norm made here results from the following considerations: If z_1 and z_2 are in K, the linear combination $\alpha_1 z_1 + \alpha_2 z_2$ is also in K since the inequality

$$(7) \qquad \| \alpha_1 z_1 + \alpha_2 z_2 \| \leq | \alpha_1 | \, \| z_1 \| + | \alpha_2 | \, \| z_2 \|$$

clearly holds; further, the product $z_1 \cdot z_2$ is also in K since the inequality

$$(8) \qquad \| z_1 z_2 \| \leq \| z_1 \| \, \| z_2 \|$$

holds. In addition, if $\dot{z}$ and z are both in K, one sees readily that $\dot{z}$ can be expressed in the form

$$(9) \qquad \dot{z} = \{i(p_1 \omega_1 + p_2 \omega_2) C_{p_1 p_2}\},$$

i.e., the series for z may be differentiated termwise to yield $\dot{z}$.

It is useful to begin by considering the linear differential equation

$$(10) \qquad \ddot{y} + c\dot{y} + y = g(t) = \{G_{p_1 p_2}\},$$

in which g is assumed to be in K, and to show that it possesses a solution in K. Such a solution is in fact given in the form

(11) $$y = Rg$$

with R an operation defined for any functions of our class (cf. (4)) by

(12) $$Rz = \{r_{p_1 p_2} C_{p_1 p_2}\}$$

when the numbers $r_{p_1 p_2}$ are given by

(13) $\quad r_{p_1 p_2} = [1 + ic(p_1\omega_1 + p_2\omega_2) - (p_1\omega_1 + p_2\omega_2)^2]^{-1}.$

One verifies readily that equation (11) does indeed furnish a solution of (10). It is important to observe that the numbers $r_{p_1 p_2}$ satisfy the inequality

(14) $$|r_{p_1 p_2}| \leq \rho,$$

with

(15) $$\begin{cases} \rho = \dfrac{1}{c\sqrt{1 - \dfrac{c^2}{4}}} & \text{if} \quad c \leq \sqrt{2}, \\[4mm] \rho = 1 & \text{if} \quad c > \sqrt{2}. \end{cases}$$

The inequality (14) is readily verified directly, or it can be inferred from the discussion at the end of Section 3 in Chapter I. It might be observed here that it is in the establishing of (14) that we make essential use of the existence of a damping term with $c > 0$. It follows that the function Rz obtained by performing the operation R defined by equation (12) has a norm which satisfies the inequality

(16) $$\|Rz\| \leq \rho \|z\|,$$

and thus belongs in K.

Instead of equation (1) it is convenient to consider the integral equation

(17) $$x = R(\beta x^2 + h),$$

which is clearly equivalent to it when the operator R is defined by (12) since we may think of (1) as written in the form $\ddot{x} + c\dot{x} + x = \beta x^2 + h(t)$. Our purpose is now to perform on (17) the iterations

(18) $$x_{n+1} = R(\beta x_n^2 + h),$$

starting with

(19) $$x_0 \equiv 0,$$

and to establish the convergence of the sequence x_n to a solution of (17) if β is taken small enough. More precisely, we prove the convergence if β satisfies the inequality

(20) $$\beta < \beta_0 = \frac{1}{4\rho^2\eta},$$

in which η and ρ are defined by (6) and (15). We also introduce the number α_0 by

(21) $$\alpha_0 = 2\rho\eta,$$

with the consequence that

(22) $$\rho(\beta\alpha^2 + \eta) < \alpha_0 \qquad \text{if } \alpha < \alpha_0, \beta < \beta_0,$$

as one can readily show. As a consequence of (22), (16), and (6) we have the inequality

(23) $$\| R(\beta z^2 + h) \| < \alpha_0$$

if $\| z \| < \alpha$; for (16) applies to $R(\beta z^2 + h)$ since this function belongs in K: our norm was in fact defined in such a way as to make this true. For the sequence x_n it therefore follows that

(24) $$\| x_n \| < \alpha_0, \qquad n = 0, 1, 2, \ldots;$$

or, in other words, the iteration process always yields functions which belong to K. From (18) and the fact that the operation R is linear it follows that

(25) $$\| x_{n+1} - x_n \| = \| R\beta(x_n^2 - x_{n-1}^2) \|,$$

and hence, in view of (8) and (16), that

(26) $$\| x_{n+1} - x_n \| \leq \rho\beta \| x_n + x_{n-1} \| \| x_n - x_{n-1} \|$$
$$\leq 2\rho\beta\alpha_0 \| x_n - x_{n-1} \|$$

by (24). From (20) and (21) we then obtain the inequality

(27) $$\| x_{n+1} - x_n \| < \frac{\beta}{\beta_0} \| x_n - x_{n-1} \|.$$

Consequently, since the class of functions K is closed, the iterations converge to a function $x(t)$ such that $\| x_n - x \| \to 0$. Furthermore, since the convergence is uniform in t it is not difficult to establish the fact that $x(t)$ satisfies the integral equation (7), and from this fact one shows easily that $x(t)$ is differentiable and is a solution of (1).

We have therefore proved the existence of combination oscillations which are in general a special type of almost periodic solutions of equation (1) under the condition (20), which may also be written as follows:

$$4\beta \| h \| < \frac{1}{\rho^2} \leq \begin{cases} c^2 \left(1 - \dfrac{c^2}{4} \right) & \text{if} \quad c \leq \sqrt{2}, \\ 1 & \text{if} \quad c > \sqrt{2}. \end{cases}$$

This condition requires either that the damping coefficient c be large enough, or that the coefficient β of the nonlinear term, or the amplitude $\| h \|$ of the excitation be small enough. Then almost periodic solutions in a neighborhood of the equilibrium state $x \equiv 0$ exist which are unique once the parameters c, β, and the function $h(t)$ in equation (1) are specified.

In the case of the Duffing equation we found, by the iteration or perturbation schemes used in Chapter IV, that in some ranges of the excitation frequency it was possible to have as many as three different solutions having the same frequency. The combination oscillations obtained above are uniquely determined due to the restriction $\| z \| \leq \alpha_0$ on the amplitude, as we have already remarked, but it would be very interesting if some other mode of attack could be invented which would lead to the analogue of the Duffing phenomena of instability and hysteresis also in the case of combination tones.

APPENDIX III

The Existence of Limit Cycles in Free Oscillations of Self-sustained Systems

1. General discussion

In Part A of Chapter V we have discussed in detail various special types of free oscillations, with particular emphasis on Rayleigh's or van der Pol's equation, for which there exists exactly one limit cycle. Our purpose in this appendix is to prove the existence of at least one limit cycle in systems for which the characteristic has certain special properties that one might expect to find normally in the applications. Later on, in Appendix VI, we shall prove that the limit cycle thus found is unique provided that an additional condition is imposed on the characteristic.

The existence of a limit cycle will be proved by carrying out a construction due to Poincaré. The idea of Poincaré, as applied to a first order system of differential equations of the form $dx/dt = P(x, v)$, $dy/dt = Q(x, v)$, is the following: A ring-shaped region in the x, v-plane free of singular points is constructed with the property that all solution curves of the system pass across the boundary of the ring into its interior when t increases. As a consequence, any solution curves which start at points inside the ring must of necessity remain inside it as t increases. One could then conclude at once, on the basis of the theorem of Poincaré and Bendixson cited in Section 13 of Chapter V (which applies since no singular points occur in the ring), that at least one limit cycle exists. However, in many cases, including those to be treated in this appendix and in the Appendices IV and VI, it is possible to prove the existence of a limit cycle rather easily without the necessity of using a theorem as general as the theorem of Poincaré and Bendixson. This is done, again following Poincaré, by showing first that all solution curves in the ring make a complete circuit around the ring as t increases, so that any solution

which starts on a cross section of the ring will cut the cross section
again for the first time for a certain larger value of t; a continuous mapping of the cross section on itself is thus set up, and the existence of
a closed solution curve or limit cycle then follows from the fact that
such a mapping can be shown to have at least one fixed point, that
is, a point which is both the initial point and the final point of a
certain solution curve that makes one complete circuit around the
ring. In all of the cases in which we use this method the cross section of the ring will be a certain segment of a straight line. It should
perhaps also be stated that the continuity of the mapping of the
segment on itself is ensured by the theorem on the continuous dependence of the solutions of the differential equations on the initial
conditions.

The fixed point theorem which is needed to justify the above
scheme for obtaining the existence of a limit cycle is easy to prove.
We give a proof of it here once for all in order to avoid interruption
in the continuity of the discussions in several places later on. Let
the mapping of the segment S: $a \leq x \leq b$ on itself be given by the
continuous function $f(x)$. We have the conditions $f(a) \geq a$ and
$f(b) \leq b$, and consequently the continuous function $f(x) - x$ is non-
negative for $x = a$ and non-positive for $x = b$ so that it vanishes some-
where in the interval $a \leq x \leq b$. In other words, a value of x exists
in the interval for which $x = f(x)$, which proves the theorem.

2. *Existence of a limit cycle*

We consider a differential equation of the form

$$(2.1) \qquad \frac{dv}{dx} = \frac{F(v) - x}{v},$$

or alternatively of the form

$$(2.1)' \qquad \begin{cases} \dfrac{dx}{dt} = v \\[2mm] \dfrac{dv}{dt} = F(v) - x \end{cases}$$

in which $F(v)$ is defined and differentiable for all v and satisfies the
following conditions:

$$(2.2) \qquad F(v) = G(v) - \alpha v, \qquad \alpha > 0,$$

(2.3) $G'(v) > 0, \qquad G'(0) > \alpha,$

(2.4) $G(-v) = -G(v),$

(2.5) $|G(v)| < c.$

This example is basically the same as the one treated by Friedrichs [40] (cf. p. 76). As was mentioned in Chapter V, the existence and uniqueness of a limit cycle for the case of the van der Pol equation have been proved by Liénard [25], and Levinson and Smith have generalized these results to a broad class of equations.

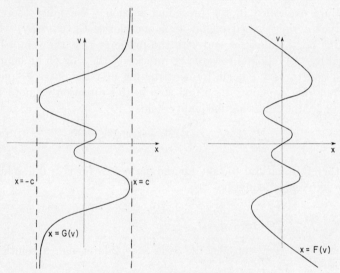

Fig. 2.1. Form of the characteristic.

The effect of these conditions is to yield a symmetric "characteristic" $x = F(v)$ of the type indicated in Figure 2.1.* It has the same qualitative appearance for large $|v|$ and for values of x and v near the origin as the characteristic in the van der Pol case: in view

---

* In practice a characteristic of the form (2.2) is to be expected with $G(v)$ having the property expressed by (2.5), since saturation effects are usually present to cause the nonlinear part $-G'(v)$ of the damping force to die out as v increases. The characteristic of the van der Pol equation does not have this property—$G(v)$ is in fact essentially v^3 in that case—but it should be recalled (cf. Section 3, Chapter V) that this form of the characteristic was taken in that case as an approximation to the actual characteristic *valid only for small values of v.*

of (2.3) the slope of the characteristic is positive at the origin, as it should be for self-sustained oscillations, while (2.2), (2.3), (2.4), and (2.5) guarantee, among other things, that the slope is negative for $|v|$ large enough since $G'(v) \to 0$ as $|v| \to \infty$. One would suspect, therefore, that a limit cycle exists. We shall indeed prove that *a limit cycle of* (2.1) *occurs under conditions* (2.2) *to* (2.5). In Appendix VI we shall add the condition that $G''(v) < 0$ for $v \neq 0$ so that only one point of inflection occurs (at the origin), and prove then that only one cycle occurs.

The existence of a limit cycle will be proved by following the procedure explained in the preceding section. We must therefore construct a ring-shaped region with boundaries C_1, C_2, say, which contains no singularities and with the property that the vector field defined by $(2.1)'$ points toward the interior of the ring on its boundaries C_1 and C_2. We begin by constructing first the outer boundary C_2 of our ring domain. As a first step in this direction we note that from (2.5), the following inequality holds:

$$(2.6) \qquad \frac{dv}{dx} < \frac{c - x - \alpha v}{v} \quad \text{for} \quad v > 0.$$

It is therefore natural to take for the part of C_2 for $v > 0$ a solution of the differential equation

$$(2.7) \qquad \frac{dv}{dx} = \frac{c - x - \alpha v}{v}.$$

If $\alpha < 2$—we shall consider the cases $\alpha \geq 2$ later on—a solution of (2.7) is

$$(2.8) \qquad \begin{cases} x = c + be^{-\alpha t/2} \left(q \cos qt + \dfrac{\alpha}{2} \sin qt \right) \\ v = -be^{-\alpha t/2} \sin qt \end{cases}$$

with q defined by

$$(2.9) \qquad q^2 = 1 - \frac{\alpha^2}{4},$$

and b an arbitrary constant. The variable t is restricted to the range $-\pi/q \leq t \leq 0$ for which $v \geq 0$ and x lies between $x = c - bqe^{\alpha\pi/2q}$ and $x = c + bq$. We determine b such that

$$(2.10) \qquad c - bqe^{\alpha\pi/2q} = -(c + bq)$$

or

(2.11) $$bq\{e^{\alpha\pi/2q} - 1\} = 2c,$$

and this can be done since $e^{\alpha\pi/2q} > 1$. The curve given by (2.8) with this value of b is defined as the curve C_2 for $v \geq 0$. We note that it crosses the x-axis for $t = 0$ and $t = -\pi/q$ at points which are symmetrical with respect to the origin. The part of C_2 for $v \leq 0$ is obtained by reflecting the part for $v \geq 0$ in the origin: clearly a simple closed curve with the origin in its interior results. Finally, the relations (2.6) and (2.7) and the symmetry of the problem guarantee that the *vector field on C_2 defined by (2.1)' points to the interior of C_2*.

For the inner boundary C_1 of the ring domain we take the circle

(2.12) $$x^2 + v^2 = r^2$$

with r chosen so small that C_1 lies entirely inside of C_2 and also so that $F(v)/v = (G(v)/v) - \alpha$ is positive for $|v| \leq r$; such a value of r exists since $G'(0) > \alpha$ by (2.3) and hence there is a value of r such that $G(v)/v > \alpha$ for $|v| \leq r$. On C_1 we have therefore, from (2.1) the inequality:

(2.13) $$\frac{dv}{dx} > -\frac{x}{v},$$

and this means that the field vector on C_1 points toward its outside, i.e. toward the inside of the ring domain bounded by C_1 and C_2. Thus the field vectors have been shown to point inward over the entire boundary of the ring. It might be added that other ways of constructing a ring domain with the desired properties could be readily given; one such method is illustrated in the paper of Levinson and Smith [24].

Consider next the intersections S_+ and S_- of the positive and negative halves of the x-axis with the ring domain between C_1 and C_2. In order to complete the desired existence proof in accordance with the scheme outlined in Section 1 we need only show that all integral curves of (2.1) which start from a point of the segment S_- to the left of the origin will eventually cross the same segment once more after making a full circuit of the ring, since the fixed point theorem for the mapping of S_- on itself then applies. To this end we first remark that any integral curve which initiates on S_- stays inside the ring domain as $t \to +\infty$ since the field vectors on the boundary of the ring point toward its interior, and in addition it enters the *upper* half plane as t

increases. Furthermore we see from the first equation of $(2.1)'$ that for $v > 0$ the x-coordinate of any solution curve increases monotonically with t; it follows that any solution curve which starts on S_- must of necessity cross S_+ for a finite value of t since no singularities occur in the ring domain and hence any solution curve leaving S_- can be continued up to the segment S_+ as t increases. Similarly, every solution curve leaving S_+ and therefore entering the lower half-plane must eventually arrive back at S_-. In this way a continuous mapping of the segment S_- on itself is defined by coordinating a given point P on S_- with the point Q at which the solution curve starting from P arrives after one circuit around the ring domain. Our existence proof is now complete upon applying the fixed point theorem to the mapping of S_- on itself.

The above proof was carried out for the case $\alpha < 2$ (cf. (2.2)), but it could easily be modified to take care of the cases $\alpha \geq 2$. All that is needed is a different construction for the curve C_2, which, however, can be carried out by integrating (2.7) explicitly and choosing an appropriate solution.

A construction similar to the one employed here has been carried out in a much more complicated three-dimensional case by Friedrichs [40]. The problem concerned an electrical circuit with a triode vacuum tube so arranged that the resulting system of differential equations was of *third order*, i.e., a system of three first order equations for three unknown functions. The existence of a limit cycle was then deduced by constructing explicitly a certain torus in the three-dimensional space, whose coordinates were the unknown functions, with the property that the field vectors at all points of the boundary surface pointed toward the interior of the torus. In addition, every integral curve was shown to make a full circuit around the torus. The fixed point theorem applied now for the mapping of a simply connected domain on itself then furnished the existence of a limit cycle.

APPENDIX IV

Relaxation Oscillations of the van der Pol Equation

In Section 6 of Chapter V we gave certain plausible geometric arguments which led us to the form of the limit of the limit cycles of the equation

$$(1) \qquad \frac{dv}{d\xi} = \epsilon^2 \frac{\left(v - \dfrac{v^3}{3}\right) - \xi}{v} = \epsilon^2 \frac{F(v) - \xi}{v}$$

when $\epsilon \to \infty$. In this appendix a rigorous proof is given (cf. Flanders and Stoker [11]) that the limit of the limit cycles is actually what was assumed in Chapter V, i.e. that it has the form indicated by the heavy line in Figure 1.

We write (1) as a system of first order equations by introducing a parameter τ:

$$(2) \qquad \begin{cases} \dfrac{dv}{d\tau} = \epsilon^2 \left(v - \dfrac{v^3}{3} - \xi\right) \\[2mm] \dfrac{d\xi}{d\tau} = v \end{cases}$$

and formulate our *theorem* as follows: 1) *For all sufficiently large ϵ the system (2) (and with it (1), of course) possesses at least one closed solution curve $\Gamma(\delta)$ containing the origin in its interior,* in which $\delta(\epsilon)$ is a quantity to be defined later.*

2) *A value of δ can always be found such that $\Gamma(\delta)$ lies in a pre-assigned but arbitrary open neighborhood of the closed curve Γ shown in Figure 1 for all sufficiently large values of ϵ.* The curve Γ is defined by

$$(3) \qquad \begin{cases} \xi = v - \dfrac{v^3}{3} & \text{for} \quad -\tfrac{2}{3} \leq \xi \leq \tfrac{2}{3}, \quad 1 \leq v \leq 2, \\[2mm] \xi = \tfrac{2}{3} & \text{for} \quad -2 \leq v \leq 1, \end{cases}$$

---

* This statement was first proved for any given ϵ by Liénard [25].

together with the corresponding curve segments symmetrical to (3) with respect to the origin.

The method of proof is as follows: As in the preceding appendix an open annular region $D(\delta)$ is first constructed in the ξ, v-plane which contains Γ in its interior, but excludes a neighborhood of the origin. $D(\delta)$ is so chosen that for all sufficiently large ϵ every integral curve of (2) through a boundary point of $D(\delta)$ passes into the *interior* of $D(\delta)$ with increasing t. It is next shown that every integral in D makes a complete circuit around the ring. From this the first conclusion of our theorem follows at once, as in Appendix III, from the fixed point theorem for the mapping of a segment on itself. The sec-

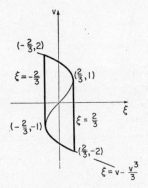

FIG. 1. The limit of limit cycles.

ond conclusion is proved by showing that for all sufficiently large ϵ it is possible to choose δ so that $D(\delta)$ is contained in an arbitrarily given open neighborhood of Γ.

The annular open set $D(\delta)$ is defined by explicit construction of its boundary curves $\bar{D}_{+}(\delta)$ and $\bar{D}_{-}(\delta)$. The curves $\bar{D}_{+}(\delta)$ and $\bar{D}_{-}(\delta)$ are shown in Figure 2, together with the curve Γ. The curve C: $\xi = v - v^3/3$ is shown by dashes. We proceed to define the curves $\bar{D}_{+}$ and $\bar{D}_{-}$, describing only half of the construction since these curves are symmetrical with respect to the origin. Symmetrical points will be denoted by such pairs as P_i, P_i' and Q_i, Q_i'. It is to be noted that the field vectors $(d\xi/d\tau, dv/d\tau)$ defined by (2) have the same symmetry property as the curves $\bar{D}_{+}$ and $\bar{D}_{-}$.

Description of $\bar{D}_+ (\delta)$:

Consider the curve C to be translated a distance 2δ in the positive ξ-direction, and let Q_1 and Q_2 be the points on the uppermost branch of the translated curve in which $\xi = -\frac{2}{3}$ and $\xi = \frac{2}{3}$ respectively.

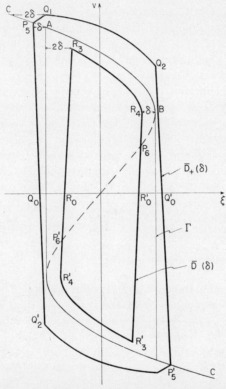

FIG. 2. The ring domain containing the limit cycle.

Let P_5 be the point on C with $\xi = -\frac{2}{3} - \delta$. P_5 and Q_1 are joined by a straight line segment. Q_2 and P_5' (the point symmetrical to P_5 with respect to the origin) are also joined by a straight line segment.   One half of the curve $\bar{D}_+(\delta)$ then consists of the segment P_5Q_1, the arc Q_1Q_2 obtained from C by a translation, and the straight line segment Q_2P_5'.

Description of $\bar{D}_-(\delta)$:

Consider C to be translated a distance δ, where δ should now be restricted to remain less than $\frac{1}{10}$ say, in the negative ξ-direction, and let R_3 and R_4 be the points on the uppermost branch of the translated curve at which $\xi = -\frac{2}{3} + 2\delta$ and $\xi = \frac{2}{3} - \delta$ respectively. Let P_6 be the point on C with coordinates, $\xi = \frac{2}{3} - \delta$, and v a value between 0 and 1 (that is, a point on the arc of C which extends from the origin to the point $(\frac{2}{3}, 1)$). We join R_4 to P_6 and P_6 to R_3' by straight line segments. One half of $\bar{D}_-(\delta)$ then consists of the arc R_3R_4 plus the two straight line segments R_4P_6 and P_6R_3'.

It is evident that Γ lies in the interior of the annular region $D(\delta)$ bounded by $\bar{D}_+(\delta)$ and $\bar{D}_-(\delta)$ and that the origin, the only singular point of (2), is not in $D(\delta)$. Since $\bar{D}_+$ and $\bar{D}_-$ would evidently both lie in an arbitrary open neighborhood of Γ when δ is appropriately chosen, our theorem will be proved once we have shown that the vectors $(d\xi/d\tau, dv/d\tau)$ at all boundary points of $D(\delta)$ point into the interior of $D(\delta)$ for all sufficiently large ϵ. We proceed to verify the correctness of this statement for both $\bar{D}_+$ and $\bar{D}_-$. Constant reference should be made to Figure 2.

The vector field on $\bar{D}_+(\delta)$:

We consider the part of $\bar{D}_+$ in the upper half plane. Let Q_0 and Q_0' be the points where $Q_2'P_5$ and Q_2P_5' cross the ξ-axis. On Q_0P_5 the slope of $\bar{D}_+$ is a negative constant, while the field vector $(d\xi/d\tau, dv/d\tau)$ in this segment slopes upward to the right, except at Q_0, where the field vector is vertically upward, as one readily sees from (2) since $d\xi/d\tau$ and $dv/d\tau$ are both non-negative. At P_5 the field vector is $(d\xi/d\tau, 0)$ with $d\xi/d\tau$ positive. Thus the field vector points inward everywhere along the closed segment Q_0P_5, since the slope of the segment Q_0P_5 is positive. With the exception of point P_5, the field vector points downward to the right on the segment P_5Q_1, since $d\xi/d\tau$ is positive while $dv/d\tau$ is negative at these points. Hence the field vectors at all points of P_5Q_1 point inward except possibly at the end-point Q_1. On the closed arc Q_1Q_2 the slope of $\bar{D}_+$ is negative and different from zero, while the field vector $(d\xi/d\tau, dv/d\tau)$ has a positive ξ-component and a negative v-component (which is bounded away from zero) and slopes down to the right more and more steeply with increase of ϵ. We can therefore for any given δ choose ϵ so large that the field vector will point toward the interior of $D(\delta)$ along the arc Q_1Q_2, with the possible exception of Q_2. Finally, the slope of the segment Q_2Q_0' is a negative constant, while the field vector along Q_2Q_0' has the same properties as along the arc Q_1Q_2. Hence one may again choose ϵ so large that the field vector will point inward along the closed segment Q_2Q_0'. On account of symmetry, the field vectors at all points of $\bar{D}_+(\delta)$ therefore point inward for any given δ once ϵ has been properly chosen.

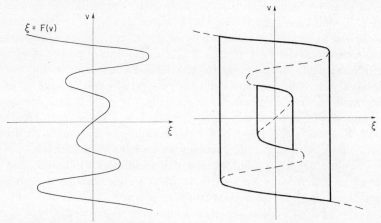

FIG. 3. A characteristic yielding more than one limit cycle.   FIG. 4. A case in which two relaxation oscillations occur.

The vector field on $\bar{D}_-(\delta)$:

Again we consider the portion of $\bar{D}_-(\delta)$ in the upper half plane. The points R_0 and R_0' are the intersections of $\bar{D}_-$ with the ξ-axis. At R_0 the field vector is directed vertically upward and thus points into the interior of $D(\delta)$. From R_0 to R_3 the field vector points upward to the right with a slope proportional to ϵ^2. By choosing ϵ sufficiently large the field vectors at all points of the closed straight line segment R_0R_3 (which has a constant positive slope) will point into the interior of $D(\delta)$ for any fixed δ. From R_3 to R_4 the slope of $\bar{D}_-(\delta)$ is negative, while the field vectors at all points of this curve segment point upward to the right; hence the field vectors point inward on this portion of $\bar{D}_-(\delta)$. The segment R_4P_6 is vertical while the field vector at all of its points is non-vertical and points toward the right, so that the field vector points inward in this segment also. Finally, the field vectors along the open segment P_6R_0' (which has a positive slope) slope downward to the right. At the upper end-point the vector is horizontal and turned to the right and at the lower end-point it is turned vertically downward. On account of symmetry, the field vectors at all points of $\bar{D}_-(\delta)$ thus point inward for any given $\delta < \frac{1}{10}$ once ϵ has been fixed appropriately.

Summing up, we see that once any small δ has been chosen a value ϵ_0 of ϵ can be found such that the field vectors at all boundary points of $D(\delta)$ point into the interior of $D(\delta)$ for all $\epsilon \geq \epsilon_0$. It is clearly sufficient to choose for ϵ_0 the largest of the values ϵ which were needed to insure that the field vectors on each of a finite number of closed curve segments should point inward. Finally we observe that every integral curve in the ring makes a complete circuit around it as τ increases since ξ increases monotonically for $v > 0$ and decreases monotonically for $v < 0$ in view of $d\xi/d\tau = v$. These facts, together with the fixed point theorem for the mapping of a segment on itself, are sufficient to prove our theorem, as we have stated earlier.

From a practical point of view it is of interest to observe that ϵ_0 must be taken large without limit when δ runs through a sequence tending to zero. This is clear from our proof.

The geometrical construction employed above would lend itself readily to generalizations of our theorem. Consider, for example, the differential equation

$$\frac{dv}{d\xi} = \epsilon^2 \frac{F(v) - \xi}{v}$$

in which $F(v)$ is one-valued and has a continuous derivative, and with $F(0) = 0$. Suppose that the curve $\xi = F(v)$ has the form indicated in Figure 3. It is not difficult to see that the same methods used above would show the existence of at least two distinct cycles for sufficiently large ϵ which would tend, as $\epsilon \to \infty$, to the closed curves indicated in Figure 4, where $\xi = F(v)$ is now indicated by a dotted curve.

APPENDIX V

The Criterion of Poincaré for Orbital Stability

In the final section of Chapter VI we discussed the *orbital stability* of periodic free vibrations of a nonlinear system by making use of the theory of Hill's equation. In this appendix we treat the same problem by a method due to Poincaré which leads to a criterion for orbital stability in a form that is of general interest. The criterion is also needed to obtain an important result in Appendix VI to follow.

We consider the following system of two first order differential equations

(1)
$$\begin{cases} \dot{x} = \dfrac{dx}{dt} = f(x, v) \\[2ex] \dot{v} = \dfrac{dv}{dt} = g(x, v) \end{cases}$$

in which f and g are functions with continuous second derivatives, say. As we know, all cases of free vibrations treated in this book are included in (1). We assume that the physical system represented by (1) has a periodic solution of period T, which means that the system (1) has a closed solution curve, or cycle $C(t)$. We assume also that no singularity of (1) occurs on C, i.e. that $\dot{x}$ and $\dot{v}$ do not vanish simultaneously on C; hence C is in particular a simple closed curve. The periodic motion represented by C is then said to be orbitally stable if the following condition is satisfied: *Any solution curve which starts from a sufficiently small neighborhood of a point on C is either a closed curve or a spiral which approaches C as $t \to +\infty$.* Otherwise the cycle C is said to be unstable. This definition is evidently in accord with the physical intuition.

We must consider a family of solutions of (1) in a neighborhood of C. It is convenient to introduce such a family C_λ depending on a parameter λ in the special way indicated in Figure 1. Without loss of generality we may assume the cycle C to be located so that the

253

x-axis falls along the normal to C (which exists because C contains no singularities) at a certain point P with coordinates $x = x_0$, $v = 0$ to which the time $t = 0$ is assigned, since our definition of stability is independent of the choice of coordinate system and the equations (1) are invariant when t is replaced by $t +$ constant. This choice of coordinate axes and initial conditions will be adhered to in all that follows. We confine our attention to solutions which initiate on the x-axis in a neighborhood $U(P)$ of P so small that any solution starting from a point in that neighborhood crosses the x-axis once more for some $t > 0$; and determine uniquely a family $C_\lambda : x(t, \lambda)$, $v(t, \lambda)$ of such solutions by the stipulation that they satisfy the initial condi-

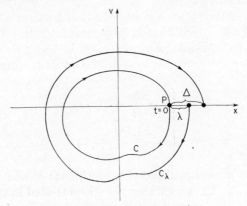

FIG. 1. Significance of quantity.

tions $x(0, \lambda) = x_0 + \lambda$, $v(0, \lambda) = 0$ for $t = 0$. The existence of such a family is ensured by the theorems on existence, uniqueness, and continuous dependence on initial conditions for solutions of ordinary differential equations; in addition $x(t, \lambda)$ and $v(t, \lambda)$ have continuous derivatives with respect to λ. The parameter λ represents the distance along the x-axis from P to the initial point of the solution curve C_λ; in particular, we have $C_0 = C$, or $x(t, 0) = x(t)$, $v(t, 0) = v(t)$. In view of our definition of stability the cycle C is therefore seen to be stable or not, depending upon whether the solutions C_λ which originate in $U(P)$ cross the x-axis for the first time (for $t > 0$) at a distance $\Delta = \Delta(\lambda)$ from P such that $\Delta \leq |\lambda|$ or $\Delta > |\lambda|$ respectively. In addition, if either of the strict inequalities $\Delta < \lambda$ or $\Delta > \lambda$ holds for all sufficiently small λ it is clear that there is a neighborhood of the cycle C within which no closed solution curves, but only spirals, occur; in this important special case we say that *the cycle C is isolated*. In Figure 1 a case of instability is shown with λ positive.

Before going on to discuss the family of solutions $C_\lambda : x(t, \lambda)$, $v(t, \lambda)$ in detail, it is useful to obtain for later purposes a result which follows from our special choice of the parameter λ and the coordinate axes. The result in question states that the determinant $W(t, \lambda)$ defined by

$$(2) \qquad W(t, \lambda) = \begin{vmatrix} x_\lambda & \dot{x} \\ v_\lambda & \dot{v} \end{vmatrix}$$

in which $x_\lambda (t, \lambda) = \partial x/\partial \lambda$, $v_\lambda (t, \lambda) = \partial v/\partial \lambda$ and $\dot{x}$ and $\dot{v}$ are, as always, the time derivatives of the same quantities, does not vanish for $t = 0$, i.e. that

$$(3) \qquad W(0, \lambda) = \dot{v}(0, \lambda) \neq 0$$

holds for all sufficiently small λ. This follows at once from the fact that λ was chosen in such a way that $\lambda = x(0, \lambda)$, $0 = v(0, \lambda)$ with the consequence that $x_\lambda = 1$, $v_\lambda = 0$ for $t = 0$, while $\dot{v}(0, \lambda)$ will be different from zero for all sufficiently small λ since $\dot{v}(0, 0)$ does not vanish by virtue of the fact that the x-axis at P is normal to C and the fact that $\dot{v}$ is continuous in λ. It is assumed from now on that λ has been so restricted.

As a preliminary step it is convenient to introduce next the following pair of linear differential equations, the "variational equations" associated with (1), for two functions δx and δv:

$$(4) \qquad \begin{cases} \delta \dot{x} = f_x \, \delta x + f_v \, \delta v \\ \delta \dot{v} = g_x \, \delta x + g_v \, \delta v, \end{cases}$$

with f_x, f_v etc. denoting the partial derivatives of the functions f and g occurring in (1). We observe that $\delta x = \dot{x}(t, \lambda)$, $\delta v = \dot{v}(t, \lambda)$ and $\delta x = x_\lambda(t, \lambda)$, $\delta v = v_\lambda(t, \lambda)$ constitute two sets of solutions of (4) if $x(t, \lambda)$ and $v(t, \lambda)$ solve the equations (1); to prove this statement, one need only insert the solution $x(t, \lambda)$, $v(t, \lambda)$ in (1) and differentiate— once with respect to t and once with respect to λ. In addition we remark that the two sets of solutions of (4) thus obtained are linearly independent since the Wronskian determinant $W(t, \lambda)$ already introduced in (2) does not vanish* because of the fact that by (3) it does not vanish for $t = 0$. The value of $W(t, \lambda)$ is determined by its value at $t = 0$ from the relation

$$(5) \qquad \log \frac{W(t, \lambda)}{W(0, \lambda)} = \int_0^t (f_x + g_v) \, dt,$$

which in turn follows from $dW/dt = (f_x + g_v)W$, as one can verify by differentiating both sides of equation (2). (Equation (5) is the analogue of equation (2.6) in Chapter VI.) Later on the relation (5) is needed only for the case $\lambda = 0$.

We turn now to the derivation of the desired criterion for stability of the cycle $x(t) = x(t, 0)$, $v(t) = v(t, 0)$ of period T. To this end it is convenient to introduce a "time shift factor" μ and set

$$(6) \qquad \begin{cases} \Delta x(t) = x(\mu t, \lambda) - x(t), \\ \Delta v(t) = v(\mu t, \lambda) - v(t). \end{cases}$$

The factor $\mu = \mu(\lambda)$ is to be chosen in such a way that

$$(7) \qquad \Delta v(T) = 0.$$

The reason for introducing the factor μ in this way is that it makes the determination of the stability or instability of a given cycle depend in a simple way on one single quantity, as follows.† The quantity $\Delta = \Delta(\lambda)$ introduced earlier (cf. Figure 1) is now the same as $\Delta x(T)$, in view of (7) and our choice of initial conditions and coordinate axes. We observe also that $\Delta x(0) = \lambda$. It is now convenient to introduce a new function $\eta = \eta(\lambda)$ through the following relations

$$(8) \qquad \Delta(\lambda) = \Delta x(T) = \eta(\lambda)\Delta x(0) = \lambda\eta(\lambda),$$

which define $\eta(\lambda)$ for $\lambda \neq 0$. We see therefore that a given cycle is stable if $|\eta| \leq 1$ and is unstable if $|\eta| > 1$ for all sufficiently small values of λ. Our object will be to show that as $\lambda \to 0$ the function $\eta(\lambda)$ approaches a value $\eta_0 > 0$, and hence that the limit cycle is stable if $\eta_0 < 1$ and unstable if $\eta_0 > 1$. The case $\eta_0 = 1$ is left undecided, and we shall not undertake to settle it later. However, if

---

* In Section 2, Chapter VI the Wronskian determinant and its relation to linear independence were discussed for a single second order linear differential equation; the similar discussion for the pair of first order equations (4) can be carried out by means of an obvious modification.

† See also the discussion in sec. 7, ch. VI, on this point.

$\eta_0 \neq 1$ we note that the given cycle is *isolated* in the sense defined above. We turn then to the problem of determining the value of η_0 .

First of all we observe that since all of the solution curves $x(t, \lambda)$, $v(t, \lambda)$ considered cross the x-axis again after $t = 0$—in fact when t is near to T—it follows that for each λ the quantity $\mu(\lambda)$ exists such that (7) holds; in addition $\mu(\lambda)$ could be shown to have a continuous second derivative, so that we may write, since obviously $\mu = 1$ for $\lambda = 0$:

$$(9) \qquad \mu = 1 + \kappa\lambda + 0(\lambda^2),$$

with κ a constant and $0(\lambda^2)$ a function of order λ^2. We also set

$$(10) \qquad \eta(\lambda) = \eta_0 + 0(\lambda).$$

The relation (7) and the relation $\Delta x(T) = \lambda\eta(\lambda)$ taken from relations (8) permit us, in view of (7), (9), and (10), to derive the following equations from (6):

$$(11) \qquad \begin{cases} v_\lambda(T, 0) + \kappa T \dot{v}(\mathrm{T}, 0) = 0 \\ x_\lambda(T, 0) + \kappa T \dot{x}(\mathrm{T}, 0) = \eta_0 \end{cases}$$

by developing in the neighborhood of $\lambda = 0$, using $v(0) = 0$, and setting $\lambda = 0$. From the equations (11) we eliminate κT to obtain for η_0 the relation

$$(12) \qquad \eta_0 = \frac{1}{\dot{v}(T, 0)} \begin{vmatrix} x_\lambda(T, 0) & \dot{x}(T, 0) \\ v_\lambda(T, 0) & \dot{v}(T, 0) \end{vmatrix} = \frac{1}{\dot{v}(T, 0)} W(T, 0),$$

the last step following from (2). The parameter λ was, however, introduced in such a way that $x_\lambda(0, 0) = 1$, $v_\lambda(0, 0) = 0$ so that $W(0, 0) = \dot{v}(0, 0) \neq 0$, according to (3). Since $v(t, 0)$ has the period T it follows that $\dot{v}(T, 0) = \dot{v}(0, 0)$ and hence we conclude from (12) that η_0 has the value

$$(13) \qquad \eta_0 = W(T, 0)/W(0, 0).$$

From relation (5) applied for $t = T$ and $\lambda = 0$ (i.e. on the cycle C) we observe that $\eta_0 > 0$ as stated above and hence we have from (13) and $\log \eta_0 = \int_0^T (f_x + g_v)\, dt$ the following criterion for stability:

$$(14) \qquad \oint_0^T (f_x + g_v)\, dt \begin{cases} < 0 & \text{stability} \\ > 0 & \text{instability} \end{cases}$$

when the line integral is evaluated along C in the sense of increasing t. This is the desired *criterion of Poincaré* for orbital stability of a given limit cycle. As remarked above, we leave aside the case in which the integral vanishes and $\eta_0 = 1$, since this case is not of direct interest to us and its treatment would require more extensive developments.

Finally we observe once more that the discussion which led to the conditions (14) shows that the cycle in question will be *isolated*, in the sense that it will have a neighborhood in which all solution curves are spirals. This observation will be used in the following Appendix VI.

APPENDIX VI

The Uniqueness of a Limit Cycle in the Free Oscillations of a Self-sustained System

1. General remarks

In Part A of Chapter V we have seen that it is quite easy to construct examples of self-sustained systems in which any number of limit cycles occur. To prove the uniqueness of the limit cycle in a given case is thus by no means an easy problem. Mention was made in Chapter V and Appendix III of the work of Liénard [25] and of Levinson and Smith [24] in proving the uniqueness of a limit cycle in certain cases. In this appendix we shall prove* the uniqueness of the limit cycle for the special class of differential equations (much less general than those considered by Levinson and Smith) treated in Appendix III, where the existence of at least one limit cycle was proved. We shall carry out the uniqueness proof by making use of the beautiful and striking idea of Levinson and Smith, which consists in deducing the uniqueness of the limit cycle from considerations of stability.

The uniqueness proof, following Levinson and Smith, is carried out in two steps: (1) First of all one derives the intuitively rather evident fact that if one isolated stable limit cycle encloses another of the same sort and there are no singularities in the ring between the two, then there exists at least one limit cycle between them. Furthermore, the set of such limit cycles is a closed set, i.e. a limit of limit cycles is itself a limit cycle. (2) One then goes on to show that *all possible limit cycles are isolated and stable* in the case under consideration. This in conjunction with the previous result obviously means that at most one stable limit cycle occurs, since the only alternative would be the existence of a limit of limit cycles.

---

* For similar considerations see Friedrichs [40].

259

2. The uniqueness proof

As we have already stated, our uniqueness proof refers to the equations (2.1)′ of Appendix III under the conditions (2.2) to (2.5) assumed there. These conditions make it possible to prove the existence of a limit cycle, but they are not sufficient to guarantee its uniqueness; we must add an additional condition. For the sake of convenience we write down the differential equations and the pre-

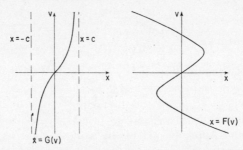

Fig. 2.1. Form of the characteristic for a unique limit cycle.

scribed conditions once more, but include also the added new condition. The differential equation is

$$(2.1)' \qquad \begin{cases} \dfrac{dx}{dt} = v, \\[2mm] \dfrac{dv}{dt} = F(v) - x. \end{cases}$$

The function $F(v)$ has a continuous second derivative and satisfies the following conditions:

$$(2.2) \qquad F(v) = G(v) - \alpha v, \qquad \alpha > 0,$$

$$(2.3) \qquad G'(v) > 0, \qquad G'(0) > \alpha,$$

$$(2.3)' \qquad G''(v) < 0, \qquad v > 0,$$

$$(2.4) \qquad G(-v) = -G(v),$$

$$(2.5) \qquad |\, G(v)\,| < c.$$

The added condition is the condition (2.3)′; obviously it ensures, in conjunction with (2.4), that $F(v)$ has exactly one point of inflexion, i.e. the origin. The characteristic $x = F(v)$, as indicated in Fig. 2.1,

now has the same qualitative appearance throughout as in the case of
the van der Pol equation, and we would suspect that only one limit cycle occurs. We shall in fact prove that *the system of equations* (2.1)′
has at most one limit cycle under the conditions (2.2) to (2.5). To do
so, we carry out, in order, the two steps of the Levinson-Smith procedure outlined at the end of the preceding section.

To prove the first part we consider the ring between two isolated
stable limit cycles. We repeat that a cycle is said to be isolated if it
has a neighborhood within which no other cycle, but only spirals,
occur; and it is stable if the spirals in that neighborhood approach it
as t increases. We know (from the discussion of Appendix III, for
example) that any such cycles contain the origin in their interiors and

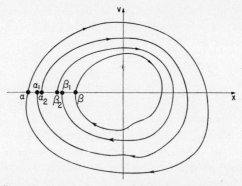

F<small>IG</small>. 2.2. Situation in which two stable isolated limit cycles occur.

that they cut the x-axis orthogonally (in view of the first equation of
(2.1)′). Just as in Appendix III we consider the segment S_-: $\alpha \leq$
$x \leq \beta$ of the negative x-axis between the two stable cycles and the
continuous mapping of S_- on itself induced by the solution curves
which start on S_- and make one circuit around the ring. We know
from Appendix IV that there is a value α_2 a little larger than α such
that the integral curve which leaves S_- at $x = \alpha_2$ with increasing t
arrives back on S_- at a point $x = \alpha_1$ with $\alpha_1 < \alpha_2$ since the cycle
through $x = \alpha$ is isolated and stable (cf. Figure 2.2). In like manner
there are two points $x = \beta_2$ and $x = \beta_1$ near $x = \beta$ with $\beta_1 > \beta_2$ with
a similar property. Consider now the segment S_-^*: $\alpha_1 \leq x \leq \beta_1$ and
the continuous mapping induced by the *backward-moving* integral
curves which make one circuit around the ring; clearly S_-^* is mapped

on *a part* of itself.   As in Appendix III this mapping has a fixed point
at $x = \gamma$, say, which corresponds to a limit cycle, and, in addition, it
is clear that the fixed point $x = \gamma$ is different from $x = \alpha$ or $x = \beta$:
actually we have $\alpha < \alpha_1 < \gamma < \beta_1 < \beta$.   This proves the *existence of
at least one limit cycle between the two given cycles*.   Furthermore, the
set of fixed points of the mapping is a closed set since the mapping
function is continuous.   It therefore follows that *a limit of limit
cycles is also a limit cycle*.

We have now to verify the correctness of the more difficult second
step in the uniqueness proof.   To this end it is necessary to make use
of the special properties of the function $F(v)$ in the differential equation $(2.1)'$ with the object of showing that the criterion of Poincaré
for stability of a limit cycle given in (14) of Appendix IV is satisfied
for any limit cycle which occurs in the present case.   In the present
case the quantity $f_x + g_v$ which figures in the criterion for stability is
$F'(v)$, as one sees from $(2.1)'$.   We have therefore to show that the
condition

$$(2.6) \qquad \oint F'(v)\, dt < 0$$

holds when the line integral is taken in the sense of increasing t around
any limit cycle C.   In accordance with a remark made at the close
of Appendix V it would then follow that the limit cycles are also isolated in the sense used above.   Our uniqueness theorem will therefore be established once the validity of the condition (2.6) is proved.

In proving the validity of (2.6) we follow essentially the same procedure as Levinson and Smith, but the argument is simpler here
because of the nature of the characteristic $x = F(v)$ defined by our
conditions.   To this end we consider the function $g(t)$ defined by

$$(2.7) \qquad g = \tfrac{1}{2}(v^2 + \dot{v}^2),$$

from which

$$(2.8) \qquad \dot{g} = v\dot{v} + \dot{v}\ddot{v} = F'(v)\dot{v}^2$$

since we assume v to satisfy $(2.1)'$.   From conditions (2.2) to (2.5) on
$F(v)$, and particularly condition $(2.3)'$, it is readily seen that there is
exactly one positive value v_0 such that

$$(2.9) \qquad F'(v) \begin{cases} > 0 & \text{for } 0 < |v| < v_0 \\ = 0 & \text{for } |v| = v_0 \\ < 0 & \text{for } |v| > v_0. \end{cases}$$

It might be noted that we use here in an essential way the symmetry of the characteristic as well as the fact that it has only one point of inflexion.

We show now that condition (2.6) will be satisfied provided that *the minimum of g on C is attained for a point with* $|v| = v_0$ *and* $\dot{v} \neq 0$, and postpone until later the proof of this fact. Assuming this, we note that $2g > v_0^2$ everywhere on C since the minimum of g satisfies this inequality by (2.7). Since $\dot{v}^2 = 2g - v^2$ from (2.7) we may write

(2.10)
$$\dot{v}^2 \begin{cases} > 2g - v_0^2 & \text{for } |v| < v_0 \\ = 2g - v_0^2 & \text{for } |v| = v_0 \\ < 2g - v_0^2 & \text{for } |v| > v_0, \end{cases}$$

from which we have, in view of (2.9):

(2.11)
$$F'(v)\dot{v}^2 > F'(v)(2g - v_0^2)$$

for all v with $|v| \neq v_0$. But since $\dot{g} = F'(v)\dot{v}^2$ we have, in view of (2.11), the inequality:

(2.12)
$$F'(v) < \frac{\dot{g}}{2g - v_0^2}, \qquad |v| \neq v_0.$$

We have seen that $2g - v_0^2 > 0$ everywhere on C. Hence we may write

(2.13)
$$\oint_c F'(v)\, dt < \oint_c \frac{\dot{g}\, dt}{2g - v_0^2}.$$

Since $(\dot{g}\, dt)/(2g - v_0^2) = \frac{1}{2}d \log (2g - v_0^2)$ is an exact differential it follows that the integral on the right-hand side has the value zero, and consequently relation (2.6) is established.

We have still to show that g attains its minimum at a point of C where $|v| = v_0$ and $\dot{v} \neq 0$. First we show that there are points on C where $|v| = v_0$, as follows: If v_0 were greater than or equal to the maximum $|v|_{max}$ of $|v|$ on C, i.e. $v_0 \geq |v|_{max}$, it would follow from (2.9) that $F'(v) \geq 0$ and hence from (2.8) that $\dot{g} \geq 0$ for all v on C. Since $\oint \dot{g}\, dt = 0$, g would necessarily be constant on C, and since this is not the case (cf. (2.3)') we conclude that $|v|_{max} > v_0$. Consequently points occur on C with $|v| = v_0$ since $v = 0$ occurs on C. We see from (2.8) and (2.9) that stationary values of g occur only for $|v| = v_0$ or $\dot{v} = 0$. Our statement will therefore follow if we can show that the minimum of g cannot occur when $\dot{v} = 0$. This can be seen as follows. First of all we observe that the points P where $\dot{v} = 0$ are the two points on C where $|v|$ is a maximum: The condition $\dot{v} = 0$ means that the point P lies on the characteristic $x = F(v)$, as we see from the second equation of $(2.1)'$. But the characteristic is cut by C at exactly two points symmetrically located with respect to the origin, since the x-coordinate on C, in view of $dx/dt = v$, changes in a monotonic way with t in each half-plane $v > 0$ or $v < 0$. Hence the minimum and maximum of v (which are numerically equal) are taken on at the points where $\dot{v} = 0$, and our first statement is proved. Next we note that $|v|_{\max} > v_0$ so that $F'(v) < 0$ at a point P (cf. (2.9)), and hence $\dot{g} \leq 0$ in the neighborhood of P (cf. (2.8)), with $\dot{g} = 0$ only at P. Thus g decreases on crossing P and hence cannot take on a minimum at P. It follows that g attains its minimum at a point on C where $|v| = v_0$, and this completes the proof of the uniqueness theorem.

BIBLIOGRAPHY

[1] Andronow, A., and A. Witt. Zur Theorie des Mitnehmens von van der Pol. Arch. für Elektrotech., 1930.

[2] Appleton, E. V. On the Anomalous Behavior of a Galvanometer. Phil. Mag., Vol. 47, 1924.

[3] Baker, J. G. Forced Vibrations with Nonlinear Spring Constants. Trans. Am. Soc. Mech. Eng., Vol. 54, 1932.

[4] Barrow, W. L. On Oscillations of a Circuit Having a Periodically Varying Capacitance. Inst. Radio Eng., 1934.

[5] Bieberbach, L. Differentialgleichungen. J. Springer, Berlin, 1930.

[6] Cartwright, M. L. Forced Oscillations in Nearly Sinusoidal Systems. Inst. Elect. Eng., Vol. 95, 1948.

[7] le Corbeiller, Ph. Les systèmes autoentretenues. Librairie scientifique, Hermann et Cie., Paris, 1931.

[8] Dorodnitsyn, A. A. Asymptotic Solution of the van der Pol Equation (Russian). Inst. Mech. of the Acad. of Sci. of the U.S.S.R., Vol. XI, 1947.

[9] Duffing, G. Erzwungene Schwingungen bei veränderlicher Eigenfrequenz. F. Vieweg u. Sohn, Braunschweig, 1918.

[10] Eckweiler, H. J. Nonlinear Differential Equations of the van der Pol Type with a Variety of Periodic Solutions. From: Studies in Non-Linear Vibration Theory, Inst. for Math. and Mech., New York Univ., 1946.

[11] Flanders, D. A., and J. J. Stoker. The Limit Case of Relaxation Oscillations. From: Studies in Non-Linear Vibration Theory, Inst. for Math. and Mech., New York Univ., 1946.

[12] Friedrichs, K. O. On Non-Linear Vibrations of Third Order. From: Studies in Non-Linear Vibration Theory, Inst. for Math. and Mech., New York Univ., 1946.

[13] Friedrichs, K. O., and J. J. Stoker. Forced Vibrations of Systems with Nonlinear Restoring Force. Quart. App. Math., Vol. I, 1943.

[14] Haag, J. Etude asymptotique des oscillations de relaxation. Ann. Sci. Ecole Norm. Sup., Vol. 60, 1943.

[15] Haag, J. Exemples concrets d'étude asymptotique d'oscillations de relaxation. Ann. Sci. Ecole Norm. Sup., Vol. 61, 1944.

[16] den Hartog, J. P. Mechanical Vibrations. McGraw-Hill Book Co., N. Y., 1940, Ch. 7, 8.

[17] Ince, E. L.   Ordinary Differential Equations.   Longmans, Green, and Co., London, 1927.

[18] John, F.   On Simple Harmonic Vibrations of a System with Non-Linear Characteristics.   From: Studies in Non-Linear Vibration Theory, Inst. for Math. and Mech., New York Univ., 1946.

[19] John, F.   On Harmonic Vibrations Out of Phase with the Exciting Force. Comm. on App. Math.   To appear.

[20] von Kármán, Th.   The Engineer Grapples with Nonlinear Problems. Bull. Am. Math. Soc., Vol. 46, 1940.

[21] Kryloff, N., and N. Bogoliuboff.   Introduction to Non-Linear Mechanics (Russian).   A free translation has been made by S. Lefschetz under this title.   Princeton Univ. Press, 1943.

[22] Lefschetz, S.   Lectures on Differential Equations.   Princeton Univ. Press, 1946.

[23] Levenson, M. E.   Harmonic and Subharmonic Response for the Duffing Equation.   Thesis, New York University, 1948.

[24] Levinson, N., and O. K. Smith.   A General Equation for Relaxation Oscillations.   Duke Math. Journal, 1942.

[25] Liénard, A.   Etude des oscillations entretenues.   Rev. Gén. d'Elect., 1928.

[26] Lindstedt, A.   Differentialgleichungen der Störungstheorie.   Mém. Acad. Sci. St. Pétersbourg, Vol. XXXI, 1883.

[27] Ludeke, C. A.   Resonance.   Journ. App. Physics, Vol. 13, 1942.

[28] Lyon, W. V., and H. E. Edgerton.   Transient Torque-Angle Characteristics of Synchronous Machines.   Trans. Am. Inst. El. Eng., Vol. 49, 1930.

[29] Martienssen, O.   Über neue Resonanzerscheinungen in Wechselstromkreisen.   Phys. Zeit., Vol. 11, 1910.

[30] Meissner, E.   Graphische Analysis vermittelst des Linienbildes einer Funktion.   Verlag der Schweiz. Bauzeitung, Zürich, 1932.

[31] Minorsky, N.   Introduction to Non-Linear Mechanics.   J. W. Edwards, Ann Arbor, Mich., 1947.

[32] van der Pol, B.   Forced Oscillations in a System with Non-linear Resistance.   Phil. Mag., 1927.

[33] Poincaré, H.   Les méthodes nouvelles de la mécanique céleste, vol. I. Gauthier-Villars, Paris, 1892.

[34] Poincaré, H.   Sur les courbes définies par une équation différentielle. Oeuvres, Gauthier-Villars, Paris, Vol. 1, 1892.

[35] Rauscher, M.   Steady Oscillations of Systems with Nonlinear and Unsymmetrical Elasticity.   Journ. App. Mech., Vol. 5, 1938.

[36] Rayleigh.   The Theory of Sound.   Dover Publications.   American Edition, 1945.

[37] Strutt, M. J. O.   Lamé-sche, Mathieu-sche und verwandte Funktionen. J. Springer, Berlin, 1932.

[38] Timoshenko, S.   Vibration Problems in Engineering.   D. von Nostrand Co., N. Y., 1937, Ch. 2.

[39] Ziegler, H.   Erzwungene Schwingungen mit konstanter Dämpfung. Ing.-Arch., Vol. IX, 1938.

[40] Friedrichs, K. O., Ph. le Corbeiller, N. Levinson, and J. J. Stoker. Non-linear Mechanics, Brown Univ., 1942–43. A set of mimeographed lecture notes.

[41] Haupt, O. Über lineare homogene Differentialgleichungen zweiter Ordnung mit periodischen Koeffizienten.   Math Ann. Bd. 79 (1919), S.278.

[42] Andronow, A. A., and C. E. Chaikin. Theory of Oscillations. English translation edited by S. Lefschetz.   Princeton Univ. Press, 1949.

INDEX